TM

ELECTRIC MOTORS & CONTROLS

2017 Edition

A Note from the Editor...
Ugly's Electric Motors and Controls is based on the 2017 *NEC®* and
is designed to be used as a quick on-the-job reference in the electrical
industry. We have tried to include the most commonly required
information in an easy-to-read format.

We salute the National Fire Protection Association for their dedication
to the sponsorship of the *National Electrical Code®*.

National Electrical Code® and *NEC®* are registered trademarks of the
National Fire Protection Association, Inc., Quincy, MA

JONES & BARTLETT
L E A R N I N G

TABLE OF CONTENTS

TABLE OF CONTENTS (continued)

TABLE OF CONTENTS (continued)

TABLE OF CONTENTS (continued)

THE LEFT-HAND RULE

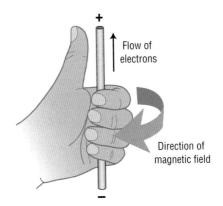

+

Flow of
electrons

Direction of
magnetic field

−

Any time an electrical current flows through a conductor, it creates a
magnetic field around it. The left-hand rule identifies the directions of
the current and the magnetic field. If you hold the conductor in your left
hand, with your thumb pointing from the negative to the positive pole,
the magnetic field around the conductor will always be in the direction
your fingers point.

MAGNETIC FIELD AROUND A COIL

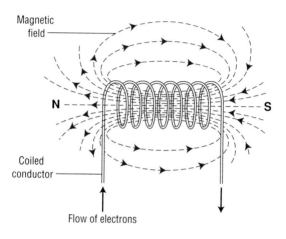

If a conductor is wound into a coil, the magnetic fields of each turn add together, producing a very strong field.

ELECTROMAGNETIC INDUCTION

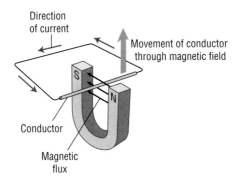

Direction
of current

Movement of conductor
through magnetic field

S

N

Conductor

Magnetic
flux

Electromagnetic induction occurs when energy is transferred in the form of magnetic fields, with no physical connection between circuits. It is the result of relative movement between a conductor and a magnetic field. Whether the conductor moves or the field moves, there is no difference. The only critical issue is that the conductor moves through the magnetic flux.

THE LEFT-HAND RULE FOR GENERATORS

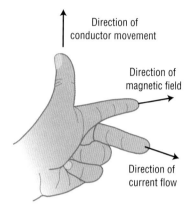

Direction of
conductor movement

Direction of
magnetic field

Direction of
current flow

A second type of left-hand rule is used to identify the motions of
conductors, currents, and fields in a generator. To do this, position your
index finger, thumb, and middle finger to be perpendicular (at 90° angles)
to one another. (See drawing.) Turn your hand so that your index finger
points in the direction of the magnetic field and your thumb in the
direction of conductor movement. Your middle finger will be pointing in
the direction of current flow.

BASIC MOTOR OPERATION

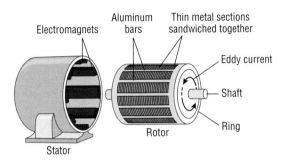

Electromagnets — Aluminum bars — Thin metal sections sandwiched together — Eddy current — Shaft — Ring — Rotor — Stator

A motor's *rotor* is built from thin metal sections so that eddy currents are reduced. These sections have embedded bars that are welded together with a ring. Current travels through the bars of the rotor.

The poles of the *stator* create a powerful magnetic field that rises and falls with each alternation of current. The rotor is either attracted or repelled by the stator poles. By cleverly arranging the positions of the poles, rotor, and current, the motor is kept moving in one direction.

BASIC DC MOTOR: SPEED CONTROL

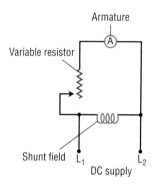

The resistor reduces the current through the armature, and with it, the speed of the motor.

CALCULATING SYNCHRONOUS SPEED

Synchronous Speed = $\dfrac{120 \times \text{Frequency}}{\text{Number of poles}}$

Example: $\dfrac{120 \times 60 \text{ Hz}}{6 \text{ poles}} = 7200 \div 6 = 1200 \text{ rpm}$

CALCULATING RUNNING SPEED

Running Speed = Synchronous Speed – Slip

Example: If slip = 5%,
running speed = synchronous speed – 5%, or
95% of synchronous speed.

$1200 \times 0.95 = 1140 \text{ rpm}$

AC MOTOR OPERATION AT OVER- AND UNDER-VOLTAGES*

	At 10% Under-Voltage
Efficiency	–2%
Temperature	+18%
Current (full load)	+12%
Speed at full load	+1%
Torque	–20%
Power factor	–4%

	At 10% Over-Voltage
Efficiency	+1%
Temperature	–9%
Current (full load)	–7%
Speed at full load	–2%
Torque	+20%
Power factor	+4%

All figures approximate.

Note that AC motors experiencing under-voltages heat up significantly. This can lead to burn-out. Lower voltages are not always less problematic.

DC MOTOR OPERATION AT OVER- AND UNDER-VOLTAGES

Characteristic	At 10% Under-Voltage		At 10% Over-Voltage	
	Shunt	Compound	Shunt	Compound
Speed	–5%	–6%	+5%	+6%
Current	+12%	+12%	–8%	–8%
Shunt Torque	–15%	–15%	+15%	+15%
Field Temp.	Increased	Decreased	Increased	Increased
Commutator Temp.	Increased	Increased	Decreased	Decreased
Armature Temp.	Increased	Increased	Decreased	Decreased

MINIMUM DEPTH OF CLEAR WORKING SPACE IN FRONT OF ELECTRICAL EQUIPMENT

Nominal Voltage to Ground	Conditions		
	1	2	3
	Minimum clear distance (feet)		
0–150	3	3	3
151–600	3	3½	4
601–2500	3	4	5
2501–9000	4	5	6
9001–25000	5	6	9
25001–75 kV	6	8	10
Above 75 kV	8	10	12

Notes:
Condition 1 = Live parts are exposed on one side of the working space only. Or: Live parts are exposed on both sides of the working space, but are all sufficiently guarded by insulating materials.
Condition 2 = Live parts are exposed on one side of the working space and grounded parts are exposed on the other side of the working space. Walls are considered to be grounded if made of concrete, brick, or tile.
Condition 3 = Live parts are exposed on both sides of the working space.

Adapted with permission from NFPA 70® 2017, National Electrical Code®, Copyright © 2016, National Fire Protection Association, Quincy, MA. This reprinted material is not the complete and official position of the NFPA on the referenced subject, which is represented only by the standard in its entirety.

Adapted from NEC® Tables 110.26(A)(1) and 110.34(A).

MINIMUM CLEARANCE OF LIVE PARTS

Nominal Voltage Rating kV	Impulse Withstand B.I.L. kV		Minimum Clearance of Live Parts, Inches			
			Phase-to-Phase		Phase-to-Ground	
	Outdoors	Indoors	Outdoors	Indoors	Outdoors	Indoors
2.4–4.16	95	60	7	4.5	6	3.0
7.2	95	75	7	5.5	6	4.0
13.8	110	95	12	7.5	7	5.0
14.4	110	110	12	9.0	7	6.5
23	150	125	15	10.5	10	7.5
34.5	150	150	15	12.5	10	9.5
	200	200	18	18.0	13	13.0
46	200	–	18	–	13	–
	250	–	21	–	17	–
69	250	–	21	–	17	–
	350	–	31	–	25	–
115	550	–	53	–	42	–
138	550	–	53	–	42	–
	650	–	63	–	50	–
161	650	–	63	–	50	–
	750	–	72	–	58	–
230	750	–	72	–	58	–
	900	–	89	–	71	–
	1050	–	105	–	83	–

- The clearances listed above are for rigid parts and bare conductors, under what the NEC calls "favorable service conditions." Where there will be conductor movement or other complications present, the above clearances should be exceeded. Extra space should always be provided if possible.
- The impulse-withstand voltage is determined by a system's surge protective equipment.

Adapted from *NEC*® Table 490.24.

HORSEPOWER RATINGS FOR NEMA STARTERS

Nema Size	Volts (VAC)	Single Phase	Three Phase	Current Rating (Amperes)
00	115	1/3	—	9
	200	—	1½	
	230	1	1½	
	460/575	—	2	
0	115	1	—	18
	200	—	3	
	230	2	3	
	460/575	—	5	
1	115	2	—	27
	200	—	7½	
	230	3	7½	
	460/575	—	10	
1P	115	3	—	36
	230	5	—	
2	115	3	—	45
	200	—	10	
	230	7½	15	
	460/575	—	25	
3	115	7½	—	90
	200	—	25	
	230	15	30	
	460/575	—	50	
4	200	—	40	135
	230	—	50	
	460/575	—	100	
5	200	—	75	270
	230	—	100	
	460/575	—	200	
6	200	—	150	540
	230	—	200	
	460/575	—	400	
7	230	—	300	90
	460/575	—	600	
8	230	—	450	1215
	460/575	—	400	
9	460/575	—	900	2250
	230	—	800	

Reprinted from Miller, Charles R. *NFPA's Pocket Electrical References, First Edition*. Jones and Bartlett Publishers.

MAXIMUM HORSEPOWER

	Three-Phase Motors											
	Full-Voltage Starting			Auto-Transformer Starting			Part-Winding Starting			Wye-Delta Starting		
Nema Size	200 VAC	230 VAC	460/575 VAC	200 VAC	230 VAC	460/575 VAC	200 VAC	230 VAC	460/575 VAC	200 VAC	230 VAC	460/575 VAC
00	1½	1½	2	—	—	—	—	—	—	—	—	—
0	3	3	5	—	—	—	—	—	—	—	—	—
1	7½	7½	10	7½	7½	10	—	10	15	10	10	15
2	10	15	25	10	15	25	20	25	40	20	25	45
3	25	30	50	25	30	50	40	50	75	45	50	75
4	40	50	100	40	50	100	70	75	150	60	75	150
5	75	100	200	75	100	200	150	150	350	150	150	300
6	150	200	400	150	200	400	—	300	600	300	350	700
7	—	300	600	—	300	600	—	450	900	500	500	1000
8	—	450	900	—	800	900	—	700	1400	750	800	1500
9	—	800	1600	—	450	1600	—	1300	2600	1500	1500	3000

Reprinted from Miller, Charles R. *NFPA's Pocket Electrical References, First Edition*. Jones and Bartlett Publishers.

RUNNING OVERLOAD UNITS

Type of Motor	Supply	Number and Location of Overload Units
1-phase AC or DC	• 2-wire • 1-phase AC or DC • ungrounded	1 overload in either conductor
1-phase AC or DC	• 2-wire • 1-phase AC or DC • one conductor ungrounded	1 overload in ungrounded conductor
1-phase AC or DC	• 3-wire • 1-phase AC or DC • grounded neutral conductor	1 overload in either ungrounded conductor
1-phase AC	• any 3-phase	1 overload in ungrounded conductor
2-phase AC	• 3-wire • 2-phase AC • ungrounded	2 overloads, one in each phase
2-phase AC	• 3-wire • 2-phase AC • one conductor grounded	2 overloads in ungrounded conductors
2-phase AC	• 4-wire • 2-phase AC • grounded or ungrounded	2 overloads, one per phase in ungrounded conductors
2-phase AC	• 5-wire • 2-phase AC • grounded neutral or ungrounded	2 overloads, one per phase in any ungrounded phase wire
3-phase AC	• any 3-phase	3 overloads, one in each phase*

* Exception: Where protected by other approved means.

Adapted from *NEC*® Table 430.37.

MOTOR BRANCH-CIRCUIT PROTECTIVE DEVICES MAXIMUM RATING OR SETTING

Type of Motor	Nontime Delay Fuse	Dual-Element (Time-Delay) Fuse	Instan-taneous Trip Breaker	Inverse Time Breaker
AC polyphase motors other than wound rotor	300% FLC	175% FLC	800% FLC	250% FLC
DC (constant voltage)	150% FLC	150% FLC	250% FLC	150% FLC
Design B energy-efficient	300% FLC	175% FLC	1100% FLC	250% FLC
Single-phase motors	300% FLC	175% FLC	800% FLC	250% FLC
Squirrel cage—other than Design B energy-efficient	300% FLC	175% FLC	800% FLC	250% FLC
Synchronous	300% FLC	175% FLC	800% FLC	250% FLC
Wound rotor	150% FLC	150% FLC	800% FLC	150% FLC

- There are exceptions to this chart. See *NEC* Sections 430.52–430.54.
- Nontime Delay Fuse = Class CC fuses.
- For nonadjustable inverse time circuit breakers, see *NEC* Section 430.52.
- For low-torque, low-speed, synchronous motors that drive reciprocating compressors, pumps, or similar loads, and which start unloaded, fuse or circuit-breaker settings need not exceed 200% of full-load current.

Adapted from *NEC*® Table 430.52.

FULL-LOAD CURRENT FOR DIRECT-CURRENT MOTORS IN AMPERES

Average DC Quantity	Armature Voltage Rating for Motors Running at Base Speed				
HP	90V	120V	180V	240V	500V
1/4	4.0	3.1	2.0	1.6	–
1/3	5.2	4.1	2.6	2.0	–
1/2	6.8	5.4	3.4	2.7	–
3/4	9.6	7.6	4.8	3.8	–
1	12.2	9.5	6.1	4.7	–
1½	–	13.2	8.3	6.6	–
2	–	17	10.8	8.5	–
3	–	25	16	12.2	–
5	–	40	27	20	–
7½	–	58	–	29	13.6
10	–	76	–	38	18
15	–	–	–	55	27
20	–	–	–	72	34
25	–	–	–	89	43
30	–	–	–	106	51
40	–	–	–	140	67
50	–	–	–	173	83
60	–	–	–	206	99
75	–	–	–	255	123
100	–	–	–	341	164
125	–	–	–	425	205
150	–	–	–	506	246
200	–	–	–	675	330

Adapted from *NEC*® Table 430.247.

**FULL-LOAD CURRENT FOR
SINGLE-PHASE ALTERNATING
CURRENT MOTORS IN AMPERES**

HP	115V	200V	208V	230V
1/4	5.8	3.3	3.2	2.9
1/3	7.2	4.1	4.0	3.6
1/2	9.8	5.6	5.4	4.9
3/4	13.8	7.9	7.6	6.9
1	16	9.2	8.8	8.0
1½	20	11.5	11	10
2	24	13.8	13.2	12
3	34	19.6	18.7	17
5	56	32.2	30.8	28
7½	80	46	44	40
10	100	57.5	55	50

The voltages listed are rated motor voltages. The listed currents are for system voltage ranges of 110 to 120 and 220 to 240.

Adapted from *NEC*® Table 430.248.

THREE-PHASE ALTERNATING CURRENT MOTORS FULL-LOAD CURRENT

HP	Induction Type Squirrel-Cage and Wound-Rotor Amperes							Synchronous Type Unity Power Factor* Amperes			
	115V	200V	208V	230V	460V	575V	2300V	230V	460V	575V	2300V
½	4.4	2.5	2.4	2.2	1.1	0.9	–	–	–	–	–
¾	6.4	3.7	3.5	3.2	1.6	1.3	–	–	–	–	–
1	8.4	4.8	4.6	4.2	2.1	1.7	–	–	–	–	–
1½	12.0	6.9	6.6	6.0	3.0	2.4	–	–	–	–	–
2	13.6	7.8	7.5	6.8	3.4	2.7	–	–	–	–	–
3	–	11.0	10.6	9.6	4.8	3.9	–	–	–	–	–
5	–	17.5	16.7	15.2	7.6	6.1	–	–	–	–	–
7½	–	25.3	24.2	22	11	9	–	–	–	–	–
10	–	32.2	30.8	28	14	11	–	–	–	–	–
15	–	48.3	46.2	42	21	17	–	–	–	–	–
20	–	62.1	59.4	54	27	22	–	–	–	–	–
25	–	78.2	74.8	68	34	27	–	53	26	21	–
30	–	92	88	80	40	32	–	63	32	26	–
40	–	120	114	104	52	41	–	83	41	33	–
50	–	150	143	130	65	52	–	104	52	42	–
60	–	177	169	154	77	62	16	123	61	49	12
75	–	221	211	192	96	77	20	155	78	62	15
100	–	285	273	248	124	99	26	202	101	81	20
125	–	359	343	312	156	125	31	253	126	101	25
150	–	414	396	360	180	144	37	302	151	121	30
200	–	552	528	480	240	192	49	400	201	161	40
250	–	–	–	–	302	242	60	–	–	–	–
300	–	–	–	–	361	289	72	–	–	–	–
350	–	–	–	–	414	336	83	–	–	–	–
400	–	–	–	–	477	382	95	–	–	–	–
450	–	–	–	–	515	412	103	–	–	–	–

The voltages listed are rated motor voltages. The currents listed are for system voltage ranges of 110 to 120, 220 to 240, 440 to 480, and 550–600 volts.
* Multiply by 1.1 and 1.25 for 90- and 80-percent power factor, respectively.

Adapted from *NEC*® Table 430.250.

FULL-LOAD CURRENT AND OTHER DATA FOR THREE-PHASE AC MOTORS

Motor Horsepower		Motor Ampere	Size Breaker	Size Starter	Heater Ampere**	Size Wire	Size Conduit
½	230V	2.2	15	00	2.530	12	¾"
	460	1.1	15	00	1.265	12	¾"
¾	230	3.2	15	00	3.680	12	¾
	460	1.6	15	00	1.840	12	¾
1	230	4.2	15	00	4.830	12	¾
	460	2.1	15	00	2.415	12	¾
1½	230	6.0	15	00	6.900	12	¾
	460	3.0	15	00	3.450	12	¾
2	230	6.8	15	0	7.820	12	¾
	460	3.4	15	00	3.910	12	¾
3	230	9.6	20	0	11.040	12	¾
	460	4.8	15	0	5.520	12	¾
5	230	15.2	30	1	17.480	12	¾
	460	7.6	15	0	8.740	12	¾
7½	230	22	45	1	25.300	10	¾
	460	11	20	1	12.650	12	¾
10	230	28	60	2	32.200	10	¾
	460	14	30	1	16.100	12	¾
15	230	42	70	2	48.300	6	1
	460	21	40	2	24.150	10	¾
20	230	54	100	3	62.100	4	1
	460	27	50	2	31.050	10	¾
25	230	68	100	3	78.200	4	1½
	460	34	50	2	39.100	8	1
30	230	80	125	3	92.000	3	1½
	460	40	70	3	46.000	8	1
40	230	104	175	4	119.600	1	1½
	460	52	100	3	59.800	6	1
50	230	130	200	4	149.500	00	2
	460	65	150	3	74.750	4	1½

* Overcurrent device may have to be increased due to starting current and load conditions. See *NEC* 430–52, Table 430–52. Wire size based on 75˚C terminations and 75˚C insulation.
** Overload heater must be based on motor nameplate and sized per *NEC* 430–32.
*** Conduit size based on Rigid Metal Conduit with some spare capacity. For minimum size and other conduit types, see *NEC* Appendix C.

19

FULL-LOAD CURRENT AND OTHER DATA FOR THREE-PHASE AC MOTORS

Motor Horsepower		Motor Ampere	Size Breaker	Size Starter	Heater Ampere**	Size Wire	Size Conduit
60	230V	154	250	5	177.10	000	2"
	460	77	200	4	88.55	3	1 ½
75	230	192	300	5	220.80	250 kcmil	2½
	460	96	200	4	110.40	1	1 ½
100	230	248	400	5	285.20	350 kcmil	3
	460	124	200	4	142.60	2/0	2
125	230	312	500	6	358.80	600 kcmil	3½
	460	156	250	5	179.40	000	2
150	230	360	600	6	414.00	700 kcmil	4
	460	180	300	5	207.00	0000	2½

Motor and Motor Circuit Conductor Protection

Motors can have starting currents three to five times (or more) than that of the motor's normal current. To allow such motors to start, the motor circuit conductors are allowed to be protected by circuit breakers and fuses at values that are higher than the actual motor and conductor ampere ratings. These larger overcurrent devices do not provide full overload protection and will open only when exposed to larger overcurrents, such as those associated with short circuits or ground faults. Overload protection (based on the actual nameplate amperes of the motor) must be used to protect the installation. This protection is usually in the form of heating elements in manual or magnetic motor starters. Small motors such as waste disposal motors have a red overload reset button built into the motor.

General Motor Rules

- Use Full-Load Current from tables instead of nameplate.
- Branch-Circuit Conductors—Use 125% of Full-Load Current to find conductor size.
- Branch-Circuit OCP Size—Use percentages given in tables for Full-Load Current.
- Feeder Conductor Size—125% of largest motor and sum of the rest.
- Feeder OCP—Use largest OCP plus rest of Full-Load Currents.

![logo] MOTOR BRANCH-CIRCUIT AND FEEDER EXAMPLE

General Motor Applications

Branch-Circuit Conductors: Use Full-Load Three-Phase Currents; From *NEC* Table 430.250, 50 HP 480 volt Three-Phase motor design B, 75 degree terminations = 65 Amperes
125% of Full-Load Current [*NEC* 430.22] 125% of 65 A = **81.25 Amperes** Conductor Selection Ampacity

Branch-Circuit Overcurrent Device: *NEC* 430.52 (C1)
(Branch-Circuit Short Circuit and Ground-Fault Protection)
Use percentages given in *NEC* 430.52 for **Type** of circuit breaker or fuse used.
50 HP 480 V 3 Ph Motor = 65 Amperes.
Nontime Fuse = 300%.
300% of 65A = 195 A. *NEC* 430.52(C1)(EX1) Next size allowed
NEC 240.6A = **200 Ampere Fuse**.

Feeder Connectors: For 50 HP and 30 HP 480 Volt Three-Phase design B motors on same feeder
Use 125% of largest full-load current and 100% of rest. (*NEC* 430.24)
50 HP 480 V 3 Ph Motor = 65A; 30 HP 480 V 3 Ph Motor = 40A
(125% of 65A) + 40A = **121.25 A** Conductor Selection Ampacity

Feeder Overcurrent Device: *NEC* 430.62(A)
(Feeder short circuit and ground-fault protection)
Use largest overcurrent protection device <u>plus</u> full-load currents of the rest of the motors.
50 HP = 200 A fuse   (65 FLC)
30 HP = 125 A fuse   (40 FLC)
200 A fuse + 40 A (FLC) = 240 A. Do not exceed this value on feeder.
Go down to a **225 A** fuse.

MAXIMUM MOTOR LOCKED-ROTOR CURRENT*

HP	115V	208V	230V	HP	115V	208V	230V
1/2	58.8	32.5	29.4	3	204	113	102
3/4	82.8	45.8	41.4	5	336	186	168
1	96	53	48	7½	480	265	240
1½	120	66	60	10	600	332	300
2	144	80	72				

Adapted from NEC® Table 430.251(A).

* Conversion Table for Selection of Disconnecting Means and Controllers as Determined from Horsepower and Voltage Rating. For use only with 430.110, 440.12, 440.41, and 455.8(C).

 MAXIMUM MOTOR LOCKED-ROTOR CURRENT IN AMPERES, TWO AND THREE-PHASE, DESIGN B, C, AND D *

HP	115V	200V	208V	230V	460V	575V
1/2	40	23	22.1	20	10	8
3/4	50	28.8	27.6	25	12.5	10
1	60	34.5	33	30	15	12
1½	80	46	44	40	20	16
2	100	57.5	55	50	25	20
3	–	73.6	71	64	32	25.6
5	–	105.8	102	92	46	36.8
7½	–	146	140	127	63.5	50.8
10	–	186.3	179	162	81	64.8
15	–	267	257	232	116	93
20	–	334	321	290	145	116
25	–	420	404	365	183	146
30	–	500	481	435	218	174
40	–	667	641	580	290	232
50	–	834	802	725	363	290
60	–	1001	962	870	435	348
75	–	1248	1200	1085	543	434
100	–	1668	1603	1450	725	580
125	–	2087	2007	1815	908	726
150	–	2496	2400	2170	1085	868
200	–	3335	3207	2900	1450	1160

Adapted from NEC® Table 430.251(B).

** Conversion Table for Selection of Disconnecting Means and Controllers as Determined from Horsepower and Voltage Rating and Design Letter. For use only with 430.110, 440.12, 440.41, and 455.8(C).

OHM'S LAW

The rate of the flow of the current is equal to electromotive force divided by resistance.

I = Intensity of Current = Amperes
E = Electromotive Force = Volts
R = Resistance = Ohms
P = Power = Watts

The three basic Ohm's Law formulas are:

$$I = \frac{E}{R} \qquad R = \frac{E}{I} \qquad E = I \times R$$

Below is a chart containing formulas derived from Ohm's Law and the electrical power formula P = E x I. To use the chart: From the center circle, select the value you wish to find: I (Amps), R (Ohms), E (Volts), or P (Watts). Then select the formula containing the values you know from the corresponding chart quadrant.

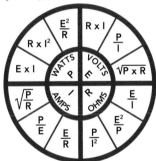

Example:
An electric appliance is rated at 1200 Watts, and is connected to 120 Volts. How much current will it draw?

$$\text{Amperes} = \frac{\text{Watts}}{\text{Volts}} \qquad I = \frac{P}{E} \qquad I = \frac{1200}{120} = 10 \text{ A}$$

What is the Resistance of the same appliance?

$$\text{Ohms} = \frac{\text{Volts}}{\text{Amperes}} \qquad R = \frac{E}{I} \qquad R = \frac{120}{10} = 12 \text{ } \Omega$$

OHM'S LAW

In the preceding example, we know the following values:

I = amps = 10		R = ohms = 12Ω	
E = volts = 120		P = watts = 1200	

We can now see how the 12 formulas in the Ohm's Law chart can be applied.

$$\text{Amps} = \sqrt{\frac{\text{Watts}}{\text{Ohms}}} \qquad I = \sqrt{\frac{P}{R}} = \sqrt{\frac{1200}{12}} = \sqrt{100} = 10A$$

$$\text{Amps} = \frac{\text{Watts}}{\text{Volts}} \qquad I = \frac{P}{E} = \frac{1200}{120} = 10A$$

$$\text{Amps} = \frac{\text{Volts}}{\text{Ohms}} \qquad I = \frac{E}{R} = \frac{120}{12} = 10A$$

$$\text{Watts} = \frac{\text{Volts}^2}{\text{Ohms}} \qquad P = \frac{E^2}{R} = \frac{120^2}{12} = \frac{14400}{12} = 1200W$$

$$\text{Watts} = \text{Volts} \times \text{Amps} \qquad P = E \times I = 120 \times 10 = 1200W$$

$$\text{Watts} = \text{Amps}^2 \times \text{Ohms} \qquad P = I^2 \times R = 100 \times 12 = 1200W$$

$$\text{Volts} = \sqrt{\text{Watts} \times \text{Ohms}} \qquad E = \sqrt{P \times R} = \sqrt{1200 \times 12} = \sqrt{14400} = 120V$$

$$\text{Volts} = \text{Amps} \times \text{Ohms} \qquad E = I \times R = 10 \times 12 = 120V$$

$$\text{Volts} = \frac{\text{Watts}}{\text{Amps}} \qquad E = \frac{P}{I} = \frac{1200}{10} = 120V$$

$$\text{Ohms} = \frac{\text{Volts}^2}{\text{Watts}} \qquad R = \frac{E^2}{P} = \frac{120^2}{1200} = \frac{14400}{1200} = 12Ω$$

$$\text{Ohms} = \frac{\text{Watts}}{\text{Amps}^2} \qquad R = \frac{P}{I^2} = \frac{1200}{100} = 12Ω$$

$$\text{Ohms} = \frac{\text{Volts}}{\text{Amps}} \qquad R = \frac{E}{I} = \frac{120}{10} = 12Ω$$

ELECTRICAL FORMULAS FOR CALCULATING AMPERES, HORSEPOWER, KILOWATTS, AND KVA

To Find	Direct Current	Alternating Current		
		Single Phase	Two-Phase Four Wire	Three Phase
Amperes when "HP" is known	$\dfrac{HP \times 746}{E \times \%EFF}$	$\dfrac{HP \times 746}{E \times \%EFF \times PF}$	$\dfrac{HP \times 746}{E \times \%EFF \times PF \times 2}$	$\dfrac{HP \times 746}{E \times \%EFF \times PF \times 1.73}$
Amperes when "KW" is known	$\dfrac{KW \times 1000}{E}$	$\dfrac{KW \times 1000}{E \times PF}$	$\dfrac{KW \times 1000}{E \times PF \times 2}$	$\dfrac{KW \times 1000}{E \times PF \times 1.73}$
Amperes when "KVA" is known		$\dfrac{KVA \times 1000}{E}$	$\dfrac{KVA \times 1000}{E \times 2}$	$\dfrac{KVA \times 1000}{E \times 1.73}$
Kilowatts (True Power)	$\dfrac{E \times I}{1000}$	$\dfrac{E \times I \times PF}{1000}$	$\dfrac{E \times I \times PF \times 2}{1000}$	$\dfrac{E \times I \times PF \times 1.73}{1000}$
Kilovolt-amperes "KVA" (Apparent Power)		$\dfrac{E \times I}{1000}$	$\dfrac{E \times I \times 2}{1000}$	$\dfrac{E \times I \times 1.73}{1000}$
Horsepower	$\dfrac{E \times I \times \%EFF}{746}$	$\dfrac{E \times I \times \%EFF \times PF}{746}$	$\dfrac{E \times I \times \%EFF \times PF \times 2}{746}$	$\dfrac{E \times I \times \%EFF \times PF \times 1.73}{746}$

Percent efficiency = % EFF = $\dfrac{\text{Output (watts)}}{\text{Input (watts)}}$ Power factor = PF = $\dfrac{\text{Power used (watts)}}{\text{Apparent power}} = \dfrac{Kw}{KVA}$

E = Volts
I = Amperes
W = Watts

Note: Direct current formulas do not use (PF, 2, or 1.73)
Single-phase formulas do not use (2 or 1.73)
Two-phase four-wire formulas do not use (1.73)
Three-phase formulas do not use (2)

25

 **TO FIND AMPERES**

Direct Current:

A.   When *Horsepower* is known:

$$\text{Amperes} = \frac{\text{Horsepower x 746}}{\text{Volts x Efficiency}} \quad \text{or} \quad I = \frac{\text{HP x 746}}{\text{E x \%EFF}}$$

What current will a travel-trailer toilet draw when equipped with a 12 volt, 1/8 HP motor, having a 96% efficiency rating?

$$I = \frac{\text{HP x 746}}{\text{E x \%EFF}} = \frac{746 \text{ x } 1/8}{12 \text{ x } 0.96} = \frac{93.25}{11.52} = 8.09 \text{ Amps}$$

B.   When *Kilowatts* are known:

$$\text{Amperes} = \frac{\text{Kilowatts x 1000}}{\text{Volts}} \quad \text{or} \quad I = \frac{\text{KW x 1000}}{\text{E}}$$

A 75 KW, 240 Volt, direct-current generator is used to power a variable-speed conveyor belt at a rock crushing plant. Determine the current.

$$I = \frac{\text{KW x 1000}}{\text{E}} = \frac{75 \text{ x } 1000}{240} = 312.5 \text{ Amps}$$

Single-Phase:

A.   When *Watts*, *Volts*, and *Power Factor* are known:

$$\text{Amperes} = \frac{\text{Watts}}{\text{Volts x Power Factor}} \quad \text{or} \quad \frac{\text{P}}{\text{E x PF}}$$

Determine the current when a circuit has a 1500 watt load, a power factor of 86%, and operates from a single-phase 240 volt source.

$$I = \frac{1500}{240 \text{ x } 0.86} = \frac{1500}{206.4} = 7.27 \text{ Amps}$$

TO FIND AMPERES

Single-Phase:

B.   When *Horsepower* is known:

$$\text{Amperes} = \frac{\text{Horsepower x 746}}{\text{Volts x Efficiency x Power Factor}}$$

Determine the amp load of a single-phase, 1/2 HP, 120 volt motor. The motor has an efficiency rating of 92%, and a power factor of 80%.

$$I = \frac{\text{HP x 746}}{\text{E x \%EFF x PF}} = \frac{1/2 \times 746}{120 \times .92 \times .80} = \frac{373}{88.32}$$

I = 4.22 Amps

C.   When *Kilowatts* are known:

$$\text{Amperes} = \frac{\text{Kilowatts x 1000}}{\text{Volts x Power Factor}} \quad \text{or} \quad I = \frac{\text{KW x 1000}}{\text{E x PF}}$$

A 240 volt single-phase circuit has a 12 KW power load, and operates at 84% power factor. Determine the current.

$$I = \frac{\text{KW x 1000}}{\text{E x PF}} = \frac{12 \times 1000}{240 \times .84} = \frac{12000}{201.6} = 59.5 \text{ Amps}$$

D.   When *Kilovolt-Ampere* is known:

$$\text{Amperes} = \frac{\text{Kilovolt-Ampere x 1000}}{\text{Volts}} \quad \text{or} \quad I = \frac{\text{KVA x 1000}}{\text{E}}$$

A 120 volt, 2 KVA, single-phase generator operating at full load will deliver 17.4 Amperes. (Prove.)

$$I = \frac{2 \times 1000}{120} = \frac{2000}{120} = 16.67 \text{ Amps}$$

Remember:
     By definition, amperes is the rate of the flow of the current.

 **TO FIND AMPERES**

Three-Phase:

A.   When *Watts, Volts,* and *Power Factor are known*:

$$\text{Amperes} = \frac{\text{Watts}}{\text{Volts x Power Factor x 1.73}}$$

or $\qquad I = \dfrac{P}{E \text{ x PF x } 1.73}$

Determine the current when a circuit has a 1500 watt load, a power factor of 86%, and operates from a three-phase, 240 volt source.

$$I = \frac{P}{E \text{ x PF x } 1.73} = \frac{1500}{240 \text{ x } .86 \text{ x } 1.73} = \frac{1500}{357.1}$$

I = 4.2 Amps

B.   When *Horsepower* is known:

$$\text{Amperes} = \frac{\text{Horsepower x 746}}{\text{Volts x Efficiency x Power Factor x 1.73}}$$

or $\qquad I = \dfrac{HP \text{ x } 746}{E \text{ x \%EFF x PF x } 1.73}$

Determine the amp load of a three-phase, 1/2 HP, 240 volt motor. The motor has an efficiency rating of 92%, and a power factor of 80%.

$$I = \frac{HP \text{ x } 746}{E \text{ x \%EFF x PF x } 1.73} = \frac{1/2 \text{ x } 746}{240 \text{ x } .92 \text{ x } .80 \text{ x } 1.73}$$

$$= \frac{373}{305.6} = 1.22 \text{ Amps}$$

TO FIND AMPERES

Three-Phase:

C. When *Kilowatts are known:*

$$\text{Amperes} = \frac{\text{Kilowatts x 1000}}{\text{Volts x Power Factor x 1.73}}$$

$$\text{or} \qquad I = \frac{\text{KW x 1000}}{\text{E x PF x 1.73}}$$

A 240 volt, three-phase circuit, has a 12 KW power load, and operates at 84% power factor. Determine the current.

$$I = \frac{\text{KW x 1000}}{\text{E x PF x 1.73}} = \frac{12000}{240 \text{ x .84 x 1.73}} = \frac{12000}{348.8}$$

$I = 34.4$ Amps

D. When *Kilovolt-Ampere* is known:

$$\text{Amperes} = \frac{\text{Kilovolt-Ampere x 1000}}{\text{E x 1.73}} = \frac{\text{KVA x 1000}}{\text{E x 1.73}}$$

A 240 volt, 4 KVA, three-phase generator operating at full load will deliver 10 Amperes. (Prove.)

$$I = \frac{\text{KVA x 1000}}{\text{E x 1.73}} = \frac{4 \text{ x 1000}}{240 \text{ x 1.73}} = \frac{4000}{415.2}$$

$I = 9.64$ Amps

TO FIND HORSEPOWER

Direct Current:

$$\text{Horsepower} = \frac{\text{Volts x Amperes x Efficiency}}{746}$$

A 12 volt motor draws a current of 8.09 amperes, and has an efficiency rating of 96%. Determine the horsepower.

$$\text{HP} = \frac{\text{E x I x \%EFF}}{746} = \frac{12 \times 8.09 \times 0.96}{746} = \frac{93.19}{746}$$

$$\text{HP} = 0.1249 = 1/8 \text{ HP}$$

Single-Phase:

$$\text{HP} = \frac{\text{Volts x Amperes x Efficiency x Power Factor}}{746}$$

A single-phase, 120 volt (AC) motor has an efficiency rating of 92%, and a power factor of 80%. Determine the horsepower if the amp load is 4.2 amperes.

$$\text{HP} = \frac{\text{E x I x \%EFF x PF}}{746} = \frac{120 \times 4.2 \times .92 \times .8}{746}$$

$$\text{HP} = \frac{370.9}{746} = .497 = 1/2 \text{ HP}$$

Three-Phase:

$$\text{HP} = \frac{\text{Volts x Amperes x Efficiency x Power Factor x 1.73}}{746}$$

A three-phase, 480 volt motor draws a current of 52 amperes. The motor has an efficiency rating of 94%, and a power factor of 80%. Determine the horsepower.

$$\text{HP} = \frac{\text{E x I x \%EFF x PF x 1.73}}{746} = \frac{480 \times 52 \times .94 \times .8 \times 1.73}{746}$$

$$\text{HP} = \frac{32472}{746} = 43.53 \text{ HP}$$

⌁ TO FIND WATTS

The electrical power in any part of a circuit is equal to the current in that part multiplied by the voltage across that part of the circuit.

A watt is the power used when one volt causes one ampere to flow in a circuit.

One horsepower is the amount of energy required to lift 33000 pounds, one foot, in one minute. The electrical equivalent of one horsepower is 745.6 watts (746 is generally used for convenience). One watt is the amount of energy required to lift 44.26 pounds, one foot, in one minute. A watt is a measurement of power—the amount of work performed in a given time.

When *Volts* and *Amperes* are known:

Power (Watts) = Volts x Amperes

A 120 volt AC circuit draws a current of 5 amperes. Determine the power consumption.

P = E x I = 120 x 5 = 600 Watts

We can now determine the resistance of this circuit.

Power = Resistance x (Amperes)²

P = R x I² or   600 = R x 25   *divide both sides of equation by 25*

$\dfrac{600}{25}$ = R   or   R = 24 Ohms

or

Power = $\dfrac{\text{(Volts)}^2}{\text{Resistance}}$ or   P = $\dfrac{E^2}{R}$ or   600 = $\dfrac{120^2}{R}$

R x 600 = 120²   or   R = $\dfrac{14400}{600}$ = 24 Ohms

Note: Refer to the formulas for Ohm's Law (as well as the formula chart) on page 24.

TO FIND KILOWATTS

Direct Current:

$$\text{Kilowatts} = \frac{\text{Volts x Amperes}}{1000}$$

A 120 volt (DC) motor draws a current of 40 amperes.
Determine the kilowatts.

$$\text{KW} = \frac{E \times I}{1000} = \frac{120 \times 40}{1000} = \frac{4800}{1000} = 4.8 \text{ KW}$$

Single-Phase:

$$\text{Kilowatts} = \frac{\text{Volts x Amperes x Power Factor}}{1000}$$

A single-phase, 120 volt (AC) motor draws a current of 20 amperes, and
has a power-factor rating of 86%. Determine the kilowatts.

$$\text{KW} = \frac{E \times I \times PF}{1000} = \frac{120 \times 20 \times .86}{1000} = \frac{2064}{1000} = 2.06 \text{ KW}$$

Three-Phase:

$$\text{Kilowatts} = \frac{\text{Volts x Amperes x Power Factor x 1.73}}{1000}$$

A three-phase, 480 volt (AC) motor draws a current of 52 amperes, and
has a power-factor rating of 80%. Determine the kilowatts.

$$\text{KW} = \frac{E \times I \times PF \times 1.73}{1000} = \frac{480 \times 52 \times .80 \times 1.73}{1000}$$

$$= \frac{34545}{1000} = 34.545 = 34.5 \text{ KW}$$

TO FIND KILOVOLT-AMPERES

Single-Phase:

$$\text{Kilovolt-Amperes} = \frac{\text{Volts} \times \text{Amperes}}{1000}$$

A single-phase, 240 volt generator delivers 41.66 amperes at full load. Determine the kilovolt-amperes rating.

$$\text{KVA} = \frac{E \times I}{1000} = \frac{240 \times 41.66}{1000} = \frac{10000}{1000} = 10 \text{ KVA}$$

Three-Phase:

$$\text{Kilovolt-Amperes} = \frac{\text{Volts} \times \text{Amperes} \times 1.73}{1000}$$

A three-phase, 480 volt generator delivers 52 amperes. Determine the kilovolt-amperes rating.

$$\text{KVA} = \frac{E \times I \times 1.73}{1000} = \frac{480 \times 52 \times 1.73}{1000} = \frac{43181}{1000}$$

$$= 43.181 = 43 \text{ KVA}$$

Note: KVA = Apparent Power = Power Before Used,
Such as the rating of a transformer.

Kirchhoff's Laws

First Law (Current):
The sum of the currents arriving at any point in a circuit must equal the sum of the currents leaving that point.

Second Law (Voltage):
The total voltage applied to any closed circuit path is always equal to the sum of the voltage drops in that path.

or

The algebraic sum of all the voltages encountered in any loop equals zero.

⎍ TO FIND INDUCTANCE

Inductance (L)

Inductance is the property of electromagnetism whereby relative motion between a conductor and a magnetic field will always induce a current in the conductor. This induced current will produce a magnetic field that opposes the original magnetic field that induced it. The effect of induction in electrical power circuits is to oppose any change in current. Inductance is produced by electrical coils, such as in transformers and motors. The unit of measurement for inductance is the henry (H).

A. To find the total inductance of coils connected in series.

$$L_T = L_1 + L_2 + L_3 + L_4$$

Determine the total inductance of four coils connected in series. Each coil has an inductance of four Henries.

$$L_T = L_1 + L_2 + L_3 + L_4$$

$$= 4 + 4 + 4 + 4 = 16 \text{ Henries}$$

B. To find the total inductance of coils connected in parallel.

$$\frac{1}{L_T} = \frac{1}{L_1} + \frac{1}{L_2} + \frac{1}{L_3} + \frac{1}{L_4}$$

Determine the total inductance of four coils connected in parallel. Each coil has an inductance of four Henries.

$$\frac{1}{L_T} = \frac{1}{L_1} + \frac{1}{L_2} + \frac{1}{L_3} + \frac{1}{L_4}$$

$$\frac{1}{L_T} = \frac{1}{4} + \frac{1}{4} + \frac{1}{4} + \frac{1}{4}$$

$$\frac{1}{L_T} = \frac{4}{4} \quad \text{OR} \quad L_T \times 4 = 1 \times 4 \quad \text{OR} \quad L_T = \frac{4}{4} = 1 \text{ Henry}$$

An induction coil is a device, consisting of two concentric coils and an interrupter, that changes a low steady voltage into a high intermittent alternating voltage by electromagnetic induction. Most often used as a spark coil.

TO FIND IMPEDANCE

Impedance (Z)

Impedance is the total opposition to an alternating current presented by a circuit. Expressed in Ohms.

A. When *Volts* and *Amperes* are known:

$$\text{Impedance} = \frac{\text{Volts}}{\text{Amperes}} \quad \text{or} \quad Z = \frac{E}{I}$$

Determine the impedance of a 120 volt AC circuit that draws a current of four amperes.

$$Z = \frac{E}{I} = \frac{120}{4} = 30 \text{ Ohms}$$

B. When *Resistance* and *Reactance* are known:

$$Z = \sqrt{\text{Resistance}^2 + \text{Reactance}^2} = \sqrt{R^2 + X^2}$$

Determine the impedance of an AC circuit when the resistance is 6 Ohms, and the reactance is 8 Ohms.

$$Z = \sqrt{R^2 + X^2} = \sqrt{36 + 64} = \sqrt{100} = 10 \text{ Ohms}$$

C. When *Resistance, Inductive Reactance,* and *Capacitive Reactance* are known:

$$Z = \sqrt{R^2 + (X_L - X_C)^2}$$

Determine the impedance of an AC circuit that has a resistance of 6 Ohms, an inductive reactance of 18 Ohms, and a capacitive reactance of 10 Ohms.

$$
\begin{aligned}
Z &= \sqrt{R^2 + (X_L - X_C)^2} \\
&= \sqrt{6^2 + (18 - 10)^2} = \sqrt{6^2 + (8)^2} \\
&= \sqrt{36 + 64} = \sqrt{100} = 10 \text{ Ohms}
\end{aligned}
$$

⎿ TO FIND REACTANCE

Reactance (X)

Reactance in a circuit is the opposition to an alternating current caused by both inductance and capacitance, but is equal only to the *difference* between capacitance and inductance. Reactance, like resistance, is measured in Ohms.

A. Inductive Reactance X_L

Inductive reactance is that element of reactance in a circuit caused by self-induction.

$$X_L = 2 \times 3.1416 \times \text{Frequency} \times \text{Inductance}$$
$$= 6.28 \qquad \times \quad F \qquad \times \quad L$$

Determine the reactance of a four-Henry coil on a 60 cycle, AC circuit.

$$X_L = 6.28 \times F \times L = 6.28 \times 60 \times 4 = 1507 \text{ Ohms}$$

B. Capacitive Reactance X_c

Capacitive reactance is that element of reactance in a circuit caused by capacitance.

$$X_c = \frac{1}{2 \times 3.1416 \times \text{Frequency} \times \text{Capacitance}}$$

$$= \frac{1}{6.28 \quad \times \quad F \quad \times \quad C}$$

Determine the reactance of a four microfarad condenser on a 60 cycle, AC circuit.

$$X_c = \frac{1}{6.28 \times F \times C} = \frac{1}{6.28 \times 60 \times .000004}$$

$$= \frac{1}{0.0015072} = 663 \text{ Ohms}$$

A Henry is a unit of inductance, equal to the inductance of a circuit in which the variation of a current at the rate of one ampere per second induces an electromotive force of one volt.

SYNCHRONOUS SPEED

Synchronous Speed (in RPM) = $\dfrac{120f}{P}$

f = frequency
P = poles per phase

 SLIP

Slip (in RPM) = Synchronous speed − actual speed

Slip percentage = $\dfrac{\text{Synchronous speed} - \text{actual speed}}{\text{Synchronous speed}} \times 100$

 LOCKED-ROTOR CURRENT THREE-PHASE

LRC = $\dfrac{1000 \times HP \times kVA/HP}{V \times 1.732 \times PF \times eff}$

LRC = Locked rotor current, in amps
HP = Horsepower
V   = Volts
PF = Power factor
eff = Motor efficiency

LOCKED-ROTOR CURRENT
SINGLE-PHASE

$$LRC = \frac{1000 \times HP \times kVA/HP}{V \times PF \times eff}$$

LRC = Locked rotor current, in amps
HP = Horsepower
V = Volts
PF = Power factor
eff = Motor efficiency

MOTOR FRAMES

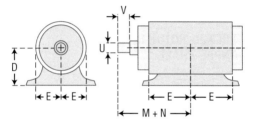

Motor frame dimensions are standardized, so that equipment and installation do not have to be customized around variable motor dimensions.

FRAME DIMENSIONS, IN INCHES

Frame No.	U	V	D	E	F	M & N	Keyway
42	3/8	–	$2^5/_8$	$1^3/_4$	27/32	$4^1/_{32}$	–
48	1/2	–	2	$2^1/_8$	$1^3/_8$	$5^3/_8$	–
56	5/8	–	$3^1/_2$	$2^7/_{16}$	$1^1/_2$	$6^1/_8$	3/16, 3/32
66	3/4	–	$4^1/_8$	$2^{15}/_{16}$	$2^1/_2$	$7^7/_8$	3/16, 3/32
143 T	7/8	2	$3^1/_2$	$2^3/_4$	2	$6^1/_2$	3/16, 3/32
145 T	7/8	2	$3^1/_2$	$2^3/_4$	$2^1/_2$	7	3/16, 3/32
182 T	$1^1/_8$	$2^1/_2$	$4^1/_2$	$3^3/_4$	$2^1/_4$	$7^3/_4$	1/4, 1/8
184 T	$1^1/_8$	$2^1/_2$	$4^1/_2$	$3^3/_4$	$2^3/_4$	$8^1/_4$	1/4, 1/8
213 T	$1^3/_8$	$3^1/_8$	$5^1/_4$	$4^1/_4$	$2^3/_4$	$9^5/_8$	5/16, 5/32
215 T	$1^3/_8$	$3^1/_8$	$5^1/_4$	$4^1/_4$	$3^1/_2$	$10^3/_8$	5/16, 5/32
254 T	$1^5/_8$	$3^3/_4$	$6^1/_4$	5	$4^1/_8$	$12^3/_8$	3/8, 3/16
256 T	$1^5/_8$	$3^3/_4$	$6^1/_4$	5	5	$13^1/_4$	3/8, 3/16
284 T	$1^7/_8$	$4^3/_8$	7	$5^1/_2$	$4^3/_4$	$14^1/_8$	1/2, 1/4
286 T	$1^7/_8$	$4^3/_8$	7	$5^1/_2$	$5^1/_2$	$14^7/_8$	1/2, 1/4
324 T	$2^1/_8$	5	8	$6^1/_4$	$5^1/_4$	$15^3/_4$	1/2, 1/4
326 T	$2^1/_8$	5	8	$6^1/_4$	6	$16^1/_2$	1/2, 1/4
364 T	$2^3/_8$	$5^5/_8$	9	7	$5^5/_8$	$17^3/_8$	5/8, 5/16
365 T	$2^3/_8$	$5^5/_8$	9	7	$6^1/_8$	$17^7/_8$	5/8, 5/16
404 T	$2^7/_8$	7	10	8	$6^1/_8$	20	3/4, 3/8
405 T	$2^7/_8$	7	10	8	$6^7/_8$	$20^3/_4$	3/4, 3/8
444 T	$3^3/_8$	$8^1/_4$	11	9	$7^1/_4$	$23^1/_4$	7/8, 7/16
445 T	$3^3/_8$	$8^1/_4$	11	9	$8^1/_4$	$24^1/_4$	7/8, 7/16

FRAME LETTER DESIGNATIONS

G	Gasoline pump motor
K	Sump pump motor
M	Oil burner motor
N	Oil burner motor
S	Short shaft or direct connection
T	Standard dimensions
U	Old designation—standard dimensions
Y	Special dimensions required
Z	Shaft extension

INSULATION CLASSES

Class	°C
A	105
B	130
F	155
H	180

Class	°F
A	221
B	266
F	311
H	356

COMMON ELECTRICAL DISTRIBUTION SYSTEMS

120/240 Volt Single-Phase Three-Wire System

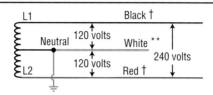

† • **Line one** ungrounded conductor colored **Black**.
† • **Line two** ungrounded conductor colored **Red**.
** • Grounded neutral conductor colored **White** or Gray.

120/240 Volt Three-Phase Four-Wire System (Delta High Leg)

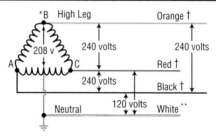

† • **A** phase ungrounded conductor colored **Black**.
†* • **B** phase ungrounded conductor colored **Orange** or tagged
   (High Leg). (Caution: 208V Orange to White)
† • **C** phase ungrounded conductor colored **Red**.
** • Grounded conductor colored **White** or Gray. (Center tap)

** Grounded conductors are required to be white or gray or to have three white or gray
stripes. See *NEC* 200.6(A).

* B phase of high leg delta must be orange or tagged.

† Ungrounded conductor colors may be other than shown; see local ordinances
or specifications.

COMMON ELECTRICAL
DISTRIBUTION SYSTEMS

120/208 Volt Three-Phase Four-Wire System (WYE Connected)

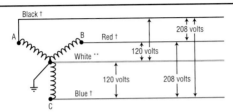

† • **A** phase ungrounded conductor colored **Black**.

† • **B** phase ungrounded conductor colored **Red**.

† • **C** phase ungrounded conductor colored **Blue**.

** Grounded conductors are required to be white or gray or to have three white or gray stripes.

277/480 Volt Three-Phase Four-Wire System (WYE Connected)

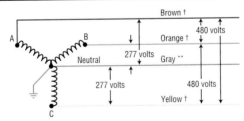

† • **A** phase ungrounded conductor colored **Brown**.

† • **B** phase ungrounded conductor colored **Orange**.

† • **C** phase ungrounded conductor colored **Yellow**.

** • Grounded neutral conductor colored **Gray**.

** Grounded conductors are required to be white or gray or to have three white stripes. See *NEC* 200.6(A).

* B phase of high leg delta must be orange or tagged.

† Ungrounded conductor colors may be other than shown; see local ordinances or specifications.

43

⌐ VOLTAGE DROP CALCULATIONS, INDUCTANCE NEGLIGIBLE

Vd = Voltage Drop
I = Current in Conductor (Amperes)
L = One-way Length of Circuit (Ft.)
Cm = Cross Section Area of Conductor (Circular Mils)
K = Resistance in ohms of one circular mil foot of conductor

K = 12.9 for Copper Conductors at 75°C
K = 21.2 for Aluminum Conductors at 75°C

Note: K value changes with temperature.
See *NEC* Chapter 9, Table 8, Notes

Single-Phase Circuits:

$$Vd = \frac{2K \times L \times I}{Cm} \qquad \text{or} \qquad {}^*Cm = \frac{2K \times L \times I}{Vd}$$

Three-Phase Circuits:

$$Vd = \frac{1.73K \times L \times I}{Cm} \qquad \text{or} \qquad {}^*Cm = \frac{1.73K \times L \times I}{Vd}$$

* *Note: Always check ampacity tables to ensure conductors' ampacity is equal to load after voltage drop calculation.*

VOLTAGE DROP CALCULATION EXAMPLES

		12 AWG	10 AWG	8 AWG	6 AWG	4 AWG	3 AWG	2 AWG	1	1/0 AWG
AMPS	VOLTS									
20	120	30	48	77	122	194	245	309	389	491
	240	60	96	154	244	388	490	618	778	982
30	120		32	51	81	129	163	206	260	327
	240		64	102	162	258	326	412	520	654
40	120			38	61	97	122	154	195	246
	240			76	122	194	244	308	390	492
50	120				49	78	98	123	156	196
	240				98	156	196	246	312	392
60	120					65	82	103	130	164
60	240					130	164	206	260	328
70	240					111	140	176	222	281
80	240						122	154	195	246
90	240							137	173	218
100	240								156	196

Distance (One Way) for 2% Voltage Drop for 120 Volts Single Phase (60° Insulation & Terminals)

See footnotes on page 46 concerning circuit load limitations.

VOLTAGE DROP CALCULATION EXAMPLES

Typical Voltage Drop Values Based on Conductor Size and One-Way Length* (60°C Termination and Insulation)

25 Feet		12 AWG	10 AWG	8 AWG	6 AWG	4 AWG	3 AWG	2 AWG	1 AWG
Amperes	20	1.98	1.24	0.78	0.49	0.31	0.25	0.19	0.15
	30		1.86	1.17	0.74	0.46	0.37	0.29	0.23
	40			1.56	0.98	0.62	0.49	0.39	0.31
	50				1.23	0.77	0.61	0.49	0.39
	60					0.93	0.74	0.58	0.46

50 Feet		12 AWG	10 AWG	8 AWG	6 AWG	4 AWG	3 AWG	2 AWG	1 AWG
Amperes	20	3.95	2.49	1.56	0.98	0.62	0.49	0.39	0.31
	30		3.73	2.34	1.47	0.93	0.74	0.58	0.46
	40			3.13	1.97	1.24	0.98	0.78	0.62
	50				2.46	1.55	1.23	0.97	0.77
	60					1.85	1.47	1.17	0.92

75 Feet		12 AWG	10 AWG	8 AWG	6 AWG	4 AWG	3 AWG	2 AWG	1 AWG
Amperes	20	5.93	3.73	2.34	1.47	0.93	0.74	0.58	0.46
	30		5.59	3.52	2.21	1.39	1.10	0.87	0.69
	40			4.69	2.95	1.85	1.47	1.17	0.92
	50				3.69	2.32	1.84	1.46	1.16
	60					2.78	2.21	1.75	1.39

100 Feet		12 AWG	10 AWG	8 AWG	6 AWG	4 AWG	3 AWG	2 AWG	1 AWG
Amperes	20	7.90	4.97	3.13	1.97	1.24	0.98	0.78	0.62
	30		7.46	4.69	2.95	1.85	1.47	1.17	0.92
	40			6.25	3.93	2.47	1.96	1.56	1.23
	50				4.92	3.09	2.45	1.94	1.54
	60					3.71	2.94	2.33	1.85

125 Feet		12 AWG	10 AWG	8 AWG	6 AWG	4 AWG	3 AWG	2 AWG	1 AWG
Amperes	20	9.88	6.21	3.91	2.46	1.55	1.23	0.97	0.77
	30		9.32	5.86	3.69	2.32	1.84	1.46	1.16
	40			7.81	4.92	3.09	2.45	1.94	1.54
	50				6.15	3.86	3.06	2.43	1.93
	60					4.64	3.68	2.92	2.31

150 Feet		12 AWG	10 AWG	8 AWG	6 AWG	4 AWG	3 AWG	2 AWG	1 AWG
Amperes	20	11.85	7.46	4.69	2.95	1.85	1.47	1.17	0.92
	30		11.18	7.03	4.42	2.78	2.21	1.75	1.39
	40			9.38	5.90	3.71	2.94	2.33	1.85
	50				7.37	4.64	3.68	2.92	2.31
	60					5.56	4.41	3.50	2.77

A two-wire 20 ampere circuit using 12 AWG with a one-way distance of 25 feet will drop 1.98 volts;

120 volts – 1.98 volts = 118.02 volts as the load voltage.

240 volts – 1.98 volts = 238.02 volts as the load voltage.

* Better economy and efficiency will result using the voltage drop method on page 44.

A continuous load cannot exceed 80% of the circuit rating.

A motor or heating load cannot exceed 80% of the circuit rating.

⏻ VOLTAGE DROP CALCULATION EXAMPLES

Single-Phase Voltage Drop

What is the voltage drop of a 240 volt single-phase circuit consisting of #8 THWN copper conductors feeding a 30 ampere load that is 150 feet in length?

Voltage Drop Formula

$$Vd = \frac{2K \times L \times I}{Cm} = \frac{2 \times 12.9 \times 150 \times 30}{16510} = \frac{116100}{16510} = 7 \text{ Volts}$$

Percentage voltage drop = 7 volts/240 volts = .029 = **2.9%**

Voltage at load = 240 volts – 7 volts = **233** V

Three-Phase Voltage Drop

What is the voltage drop of a 480 volt three-phase circuit consisting of 250 kcmil THWN copper conductors that supply a 250 ampere load that is 500 feet from the source?

Voltage Drop Formula

250 kcmil = 250000 circular mils

$$Vd = \frac{1.73K \times L \times I}{Cm} = \frac{1.73 \times 12.9 \times 500 \times 250}{250000} = \frac{2789625}{250000} = 11 \text{ Volts}$$

Percentage voltage drop = 11 volts/480 volts = .0229 = **2.29%**

Voltage at load = 480 volts – 11 volts = **469** volts

Note: Always check ampacity tables for conductors selected.

MAXIMUM PERMISSIBLE CAPACITOR KVAR

For Use with Open-Type Three-Phase Sixty-Cycle Induction Motors

Motor Rating HP	3600 RPM		1800 RPM		1200 RPM	
	Maximum Capacitor Rating KVAR	Reduction in Line Current %	Maximum Capacitor Rating KVAR	Reduction in Line Current %	Maximum Capacitor Rating KVAR	Reduction in Line Current %
10	3	10	3	11	3.5	14
15	4	9	4	10	5	13
20	5	9	5	10	6.5	12
25	6	9	6	10	7.5	11
30	7	8	7	9	9	11
40	9	8	9	9	11	10
50	12	8	11	9	13	10
60	14	8	14	8	15	10
75	17	8	16	8	18	10
100	22	8	21	8	25	9
125	27	8	26	8	30	9
150	32.5	8	30	8	35	9
200	40	8	37.5	8	42.5	9

Motor Rating HP	900 RPM		720 RPM		600 RPM	
	Maximum Capacitor Rating KVAR	Reduction in Line Current %	Maximum Capacitor Rating KVAR	Reduction in Line Current %	Maximum Capacitor Rating KVAR	Reduction in Line Current %
10	5	21	6.5	27	7.5	31
15	6.5	18	8	23	9.5	27
20	7.5	16	9	21	12	25
25	9	15	11	20	14	23
30	10	14	12	18	16	22
40	12	13	15	16	20	20
50	15	12	19	15	24	19
60	18	11	22	15	27	19
75	21	10	26	14	32.5	18
100	27	10	32.5	13	40	17
125	32.5	10	40	13	47.5	16
150	37.5	10	47.5	12	52.5	15
200	47.5	10	60	12	65	14

Note: If capacitors of a lower rating than the values given in the table are used, the percentage reduction in line current given in the table should be reduced proportionately.

48

POWER-FACTOR CORRECTION

	Total Values × KW of Capacitors Needed to Correct From Existing to Desired Power Factor					
Existing Power Factor %	Corrected Power Factor					
	100%	95%	90%	85%	80%	75%
50	1.732	1.403	1.247	1.112	0.982	0.850
52	1.643	1.314	1.158	1.023	0.893	0.761
54	1.558	1.229	1.073	0.938	0.808	0.676
55	1.518	1.189	1.033	0.898	0.768	0.636
56	1.479	1.150	0.994	0.859	0.729	0.597
58	1.404	1.075	0.919	0.784	0.654	0.522
60	1.333	1.004	0.848	0.713	0.583	0.451
62	1.265	0.936	0.780	0.645	0.515	0.383
64	1.201	0.872	0.716	0.581	0.451	0.319
65	1.168	0.839	0.683	0.548	0.418	0.286
66	1.139	0.810	0.654	0.519	0.389	0.257
68	1.078	0.749	0.593	0.458	0.328	0.196
70	1.020	0.691	0.535	0.400	0.270	0.138
72	0.964	0.635	0.479	0.344	0.214	0.082
74	0.909	0.580	0.424	0.289	0.159	0.027
75	0.882	0.553	0.397	0.262	0.132	
76	0.855	0.526	0.370	0.235	0.105	
78	0.802	0.473	0.317	0.182	0.052	
80	0.750	0.421	0.265	0.130		
82	0.698	0.369	0.213	0.078		
84	0.646	0.317	0.161			
85	0.620	0.291	0.135			
86	0.594	0.265	0.109			
88	0.540	0.211	0.055			
90	0.485	0.156				
92	0.426	0.097				
94	0.363	0.034				
95	0.329					

Typical Problem: With a load of 500 KW at 70% power factor, it is desired to find the KVA of capacitors required to correct the power factor to 85%.

Solution: From the table, select the multiplying factor 0.400 corresponding to the existing 70%, and the corrected 85% power factor. 0.400 x 500 = 200 KVA of capacitors required.

POWER FACTOR AND EFFICIENCY EXAMPLE

A squirrel-cage induction motor is rated 10 horsepower, 208 volt, three phase and has a nameplate rating of 27.79 amperes. A wattmeter reading indicates 8 kilowatts of consumed (true) power. Calculate apparent power (KVA), power factor, efficiency, internal losses, and size the capacitor in kilovolts reactive (KVAR) needed to correct the power factor to unity (100%).

Apparent input power: kilovolt-amperes (KVA)

$KVA = (E \times I \times 1.73)/1000 = (208 \times 27.79 \times 1.73)/1000 =$ **10 KVA**

Power factor (PF) = ratio of true power (KW) to apparent power (KVA). Kilowatts / kilovolt-amperes = 8 KW/10 KVA = .8 = **80% Power Factor** 80% of the 10-KVA apparent power input performs work.

Motor output in kilowatts = 10 horsepower × 746 watts = 7460 watts = **7.46 KW.**

Efficiency = watts out/watts in = 7.46 KW/8 KW = .9325 = **93.25% Efficiency.**

Internal losses (heat, friction, hysteresis) = 8 KW – 7.46 KW = **.54 KW** (540 watts)

Kilovolt-amperes reactive (KVAR)   (Power stored in motor magnetic field)

$KVAR = \sqrt{KVA^2 - KW^2} = \sqrt{10\ KVA^2 - 8\ KW^2} = \sqrt{100 - 64} = \sqrt{36}$ = **6 KVAR**

The size capacitor needed to equal the motor's stored reactive power is 6 KVAR. (A capacitor stores reactive power in its electrostatic field).

⚡ POWER FACTOR AND EFFICIENCY EXAMPLE

The power source must supply the current to perform work and maintain the motor's magnetic field. Before power factor correction, this was 27.79 amperes. The motor magnetizing current after power factor correction is supplied by circulation of current between the motor and the electrostatic field of the capacitor and is no longer supplied by power source after initial startup.

The motor feeder current after correction to 100% will equal the amount required by the input watts in this case (8 KW × 1000)/(208 volts × 1.73) = **22.23 amps.**

- Kilo = (1000 example: 1000 watts = 1 kilowatt)

- Inductive loads (motors, coils) have lagging currents, and capacitive loads have leading currents.

- Inductance and capacitance have opposite effects in a circuit and can cancel each other out.

LOCKED-ROTOR CODE LETTERS

Letter Code	Kilovolt-Ampere per Horsepower with Locked Rotor	Letter Code	Kilovolt-Ampere per Horsepower with Locked Rotor
A	0–3.14	L	9.0–9.99
B	3.15–3.54	M	10.0–11.19
C	3.55–3.99	N	11.2–12.49
D	4.0–4.49	P	12.5–13.99
E	4.5–4.99	R	14.0–15.99
F	5.0–5.59	S	16.0–17.99
G	5.6–6.29	T	18.0–19.99
H	6.3–7.09	U	20.0–22.39
J	7.1–7.99	V	22.4 and up
K	8.0–8.99		

The *National Electrical Code*® requires that all alternating current motors rated 1/2 horsepower or more (except for polyphase wound-rotor motors) must have code letters on their nameplates indicating motor input with locked rotor (in kilovolt-amperes per horsepower). If you know the horsepower and voltage rating of a motor and its "Locked KVA per Horsepower" (from above table), you can calculate the locked-rotor current using the following formulas.

LOCKED-ROTOR CODE LETTERS

Single-Phase Motors

$$\text{Locked-Rotor Current} = \frac{HP \times KVA_{hp} \times 1000}{E}$$

Three-Phase Motors

$$\text{Locked-Rotor Current} = \frac{HP \times KVA_{hp} \times 1000}{E \times 1.73}$$

Example: What is the maximum locked-rotor current for a 480 volt 25 horsepower code letter F motor?
(from the table on page 52, code letter F = 5.59 KVA_{hp})

$$I = \frac{HP \times KVA_{hp} \times 1000}{E \times 1.73} = \frac{25 \times 5.59 \times 1000}{480 \times 1.73} = \textbf{168.29} \text{ Amperes}$$

THREE-PHASE AC MOTOR WINDINGS AND CONNECTIONS

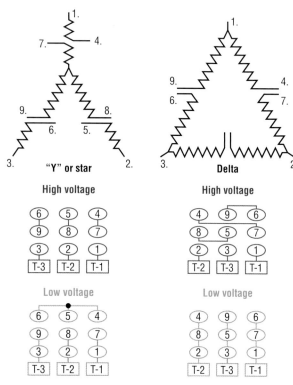

"Y" or star

Delta

High voltage

High voltage

Low voltage

Low voltage

Notes:
1. The most important part of any motor is the nameplate. Check the data given on the plate before making the connections.
2. To change rotation direction of three-phase motor, swap any 2 T-leads.

NEMA ENCLOSURE TYPES
NONHAZARDOUS LOCATIONS

The following are definitions of NEMA Enclosure Types. For more detailed and complete information, see NEMA Standards Publication 250-2014, "Enclosures for Electrical Equipment (1000 Volts Maximum)."

In **Nonhazardous Locations**, the specific enclosure types, their applications, and the environmental conditions they are designed to protect against, when completely and properly installed, are as follows:

Type 1—Enclosures constructed for indoor use to provide a degree of protection to personnel against incidental contact with the enclosed equipment and to provide a degree of protection against falling dirt.

Type 2—Enclosures constructed for indoor use to provide a degree of protection to personnel against incidental contact with the enclosed equipment, to provide a degree of protection against falling dirt, and to provide a degree of protection against dripping and light splashing of liquids.

Type 3—Enclosures constructed for either indoor or outdoor use to provide a degree of protection to personnel against incidental contact with the enclosed equipment; to provide a degree of protection against falling dirt, rain, sleet, snow, and windblown dust; and that will be undamaged by the external formation of ice on the enclosure.

Type 3R—Enclosures constructed for either indoor or outdoor use to provide a degree of protection to personnel against incidental contact with the enclosed equipment; to provide a degree of protection against falling dirt, rain, sleet, and snow; and that will be undamaged by the external formation of ice on the enclosure.

Type 3S—Enclosures constructed for either indoor or outdoor use to provide a degree of protection to personnel against incidental contact with the enclosed equipment; to provide a degree of protection against falling dirt, rain, sleet, snow, and windblown dust; and in which the external mechanism(s) remain operable when ice laden.

Reprinted from NEMA 250-2014 *by permission of the National Electrical Manufacturers Association.*

NEMA ENCLOSURE TYPES
NONHAZARDOUS LOCATIONS

Type 3X—Enclosures constructed for either indoor or outdoor use to provide a degree of protection to personnel against access to hazardous parts; to provide a degree of protection of the equipment inside the enclosure against ingress of solid foreign objects (falling dirt and windblown dust); to provide a degree of protection with respect to harmful effects on the equipment due to the ingress of water (rain, sleet, snow); that provides an increased level of protection against corrosion and that will be undamaged by the external formation of ice on the enclosure.

Type 3RX—Enclosures constructed for either indoor or outdoor use to provide a degree of protection to personnel against access to hazardous parts; to provide a degree of protection of the equipment inside the enclosure against ingress of solid foreign objects (falling dirt); to provide a degree of protection with respect to harmful effects on the equipment due to the ingress of water (rain, sleet, snow); that will be undamaged by the external formation of ice on the enclosure that provides an increased level of protection against corrosion.

Type 3SX—Enclosures constructed for either indoor or outdoor use to provide a degree of protection to personnel against access to hazardous parts; to provide a degree of protection of the equipment inside the enclosure against ingress of solid foreign objects (falling dirt and windblown dust); to provide a degree of protection with respect to harmful effects on the equipment due to the ingress of water (rain, sleet, snow); that provides an increased level of protection against corrosion; and for which the external mechanism(s) remain(s) operable when ice laden.

Type 4—Enclosures constructed for either indoor or outdoor use to provide a degree of protection to personnel against incidental contact with the enclosed equipment; to provide a degree of protection against falling dirt, rain, sleet, snow, windblown dust, splashing water, and hose-directed water; and that will be undamaged by the external formation of ice on the enclosure.

Reprinted from NEMA 250-2014 *by permission of the National Electrical Manufacturers Association.*

NEMA ENCLOSURE TYPES
NONHAZARDOUS LOCATIONS

Type 4X—Enclosures constructed for either indoor or outdoor use to provide a degree of protection to personnel against incidental contact with the enclosed equipment; to provide a degree of protection against falling dirt, rain, sleet, snow, windblown dust, splashing water, hose-directed water, and corrosion; and that will be undamaged by the external formation of ice on the enclosure.

Type 5—Enclosures constructed for indoor use to provide a degree of protection to personnel against incidental contact with the enclosed equipment; to provide a degree of protection against falling dirt; against settling airborne dust, lint, fibers, and flyings; and to provide a degree of protection against dripping and light splashing of liquids.

Type 6—Enclosures constructed for either indoor or outdoor use to provide a degree of protection to personnel against incidental contact with the enclosed equipment; to provide a degree of protection against falling dirt; against hose-directed water and the entry of water during occasional temporary submersion at a limited depth; and that will be undamaged by the external formation of ice on the enclosure.

Type 6P—Enclosures constructed for either indoor or outdoor use to provide a degree of protection to personnel against incidental contact with the enclosed equipment; to provide a degree of protection against falling dirt; against hose-directed water and the entry of water during prolonged submersion at a limited depth; and that will be undamaged by the external formation of ice on the enclosure.

Type 12—Enclosures constructed (without knockouts) for indoor use to provide a degree of protection to personnel against incidental contact with the enclosed equipment; to provide a degree of protection against falling dirt; against circulating dust, lint, fibers, and flyings; and against dripping and light splashing of liquids.

Reprinted from NEMA 250-2014 *by permission of the National Electrical Manufacturers Association.*

NEMA ENCLOSURE TYPES
NONHAZARDOUS LOCATIONS

Type 12K—Enclosures constructed (with knockouts) for indoor use to provide a degree of protection to personnel against incidental contact with the enclosed equipment; to provide a degree of protection against falling dirt; against circulating dust, lint, fibers, and flyings; and against dripping and light splashing of liquids.

Type 13—Enclosures constructed for indoor use to provide a degree of protection to personnel against incidental contact with the enclosed equipment; to provide a degree of protection against falling dirt; against circulating dust, lint, fibers, and flyings; and against the spraying, splashing, and water, oil, and noncorrosive coolants.

NEMA ENCLOSURE TYPES
HAZARDOUS LOCATIONS

In hazardous locations, when completely and properly installed and maintained, Type 7 and Type 10 enclosures are designed to contain an internal explosion without causing an external hazard. Type 8 enclosures are designed to prevent combustion through the use of oil-immersed equipment. Type 9 enclosures are designed to prevent the ignition of combustible dust.

Type 7—Enclosures constructed for indoor use in hazardous locations classified as Class I, Division 1, Groups A, B, C, or D as defined in *NFPA 70.*

Type 8—Enclosures constructed for either indoor or outdoor use in hazardous locations classified as Class I, Division 1, Groups A, B, C, or D as defined in *NFPA 70.*

Type 9—Enclosures constructed for indoor use in hazardous conditions classified as Class II, Division 1, Groups E, F, or G as defined in *NFPA 70.*

Type 10—Enclosures constructed to meet the requirements of the Mine Safety and Health Administration, 30 CFR, Part 18.

Reprinted from NEMA 250-2014 *by permission of the National Electrical Manufacturers Association.*

U.S. WEIGHTS AND MEASURES

Linear Measures

		1 Inch	= 2.540 Centimeters
12	Inches	= 1 Foot	= 3.048 Decimeters
3	Feet	= 1 Yard	= 9.144 Decimeters
5.5	Yards	= 1 Rod	= 5.029 Meters
40	Rods	= 1 Furlong	= 2.018 Hectometers
8	Furlongs	= 1 Mile	= 1.609 Kilometers

Mile Measurements

1 Statute Mile	=	5280	Feet
1 Scots Mile	=	5952	Feet
1 Irish Mile	=	6720	Feet
1 Russian Verst	=	3504	Feet
1 Italian Mile	=	4401	Feet
1 Spanish Mile	=	15084	Feet

Other Linear Measurements

1 Hand	=	4 Inches	1 Link	=	7.92	Inches
1 Span	=	9 Inches	1 Fathom	=	6	Feet
1 Chain	=	22 Yards	1 Furlong	=	10	Chains
			1 Cable	=	608	Feet

Square Measures

144	Square Inches	= 1	Square Foot
9	Square Feet	= 1	Square Yard
30 1/4	Square Yards	= 1	Square Rod
40	Rods	= 1	Rood
4	Roods	= 1	Acre
640	Acres	= 1	Square Mile
1	Square Mile	= 1	Section
36	Sections	= 1	Township

U.S. WEIGHTS AND MEASURES

Cubic or Solid Measures

1 Cu. Foot	=	1728	Cu. Inches
1 Cu. Yard	=	27	Cu. Feet
1 Cu. Foot	=	7.48	Gallons
1 Gallon (Water)	=	8.34	Lbs.
1 Gallon (U.S.)	=	231	Cu. Inches of Water
1 Gallon (Imperial)	=	$277\frac{1}{4}$	Cu. Inches of Water

Liquid Measurements

1 Pint	=	4 Gills
1 Quart	=	2 Pints
1 Gallon	=	4 Quarts
1 Firkin	=	9 Gallons (Ale or Beer)
1 Barrel	=	42 Gallons (Petroleum or Crude Oil)

Dry Measures

1 Quart	= 2 Pints
1 Peck	= 8 Quarts
1 Bushel	= 4 Pecks

U.S. WEIGHTS AND MEASURES

Weight Measurements (Mass)

A. Avoirdupois Weight:

1 Ounce	=	16 Drams
1 Pound	=	16 Ounces
1 Hundredweight	=	100 Pounds
1 Ton	=	2000 Pounds

B. Troy Weight:

1 Carat	=	3.17 Grains
1 Pennyweight	=	20 Grains
1 Ounce	=	20 Pennyweights
1 Pound	=	12 Ounces
1 Long Hundred-Weight	=	112 Pounds
1 Long Ton	=	20 Long Hundredweights
	=	2240 Pounds

C. Apothecaries Weight:

1 Scruple	= 20 Grains	=	1.296	Grams
1 Dram	= 3 Scruples	=	3.888	Grams
1 Ounce	= 8 Drams	=	31.1035	Grams
1 Pound	= 12 Ounces	=	373.2420	Grams

D. Kitchen Weights and Measures:

1 U.S. Pint	= 16	Fl. Ounces
1 Standard Cup	= 8	Fl. Ounces
1 Tablespoon	= 0.5	Fl. Ounces (15 Cu. Cms.)
1 Teaspoon	= 0.16	Fl. Ounces (5 Cu. Cms.)

METRIC SYSTEM

Prefixes

A.	Mega	= 1000000		E.	Deci	= 0.1
B.	Kilo	= 1000		F.	Centi	= 0.01
C.	Hecto	= 100		G.	Milli	= 0.001
D.	Deka	= 10		H.	Micro	= 0.000001

Linear Measures

(The Unit is the Meter = 39.37 Inches)

1 Centimeter	= 10	Millimeters	=	0.3937011	In.
1 Decimeter	= 10	Centimeters	=	3.9370113	Ins.
1 Meter	= 10	Decimeters	=	1.0936143	Yds.
			=	3.2808429	Ft.
1 Dekameter	= 10	Meters	=	10.936143	Yds.
1 Hectometer	= 10	Dekameters	=	109.36143	Yds.
1 Kilometer	= 10	Hectometers	=	0.62137	Mile
1 Myriameter	= 10000 Meters				

Square Measures

(The Unit is the Square Meter = 1549.9969 Sq. Inches)

1 Sq. Centimeter	= 100 Sq. Millimeters	=	0.1550	Sq. In.
1 Sq. Decimeter	= 100 Sq. Centimeters	=	15.550	Sq. Ins.
1 Sq. Meter	= 100 Sq. Decimeters	=	10.7639	Sq. Ft.
1 Sq. Dekameter	= 100 Sq. Meters	=	119.60	Sq. Yds.
1 Sq. Hectometer	= 100 Sq. Dekameters			
1 Sq. Kilometer	= 100 Sq. Hectometers			

(The Unit is the "Are" = 100 Sq. Meters)

1 Centiare	= 10 Milliares	=	10.7643	Sq. Ft.
1 Deciare	= 10 Centiares	=	11.96033	Sq. Yds.
1 Are	= 10 Deciares	=	119.6033	Sq. Yds.
1 Dekare	= 10 Ares	=	0.247110	Acres
1 Hektare	= 10 Dekares	=	2.471098	Acres
1 Sq. Kilometer	= 100 Hektares	=	0.38611	Sq. Mile

 **METRIC SYSTEM**

Cubic Measures

(The Unit is the "Stere" = 61025.38659 CU. INS.)

1 Decistere	= 10 Centisteres	= 3.531562	Cu. Ft.
1 Stere	= 10 Decisteres	= 1.307986	Cu. Yds.
1 Dekastere	= 10 Steres	= 13.07986	Cu. Yds.

(The Unit is the Meter = 39.37 Inches)

1 Cu. Centimeter	= 1000 Cu. Millimeters	= 0.06102	Cu. In.
1 Cu. Decimeter	= 1000 Cu. Centimeters	= 61.02374	Cu. In.
1 Cu. Meter	= 1000 Cu. Decimeters	= 35.31467	Cu. Ft.
	= 1 Stere	= 1.30795	Cu. Yds.
1 Cu. Centimeter (Water)		= 1 Gram	
1000 Cu. Centimeters (Water) = 1 Liter		= 1 Kilogram	
1 Cu. Meter (1000 Liters)		= 1 Metric Ton	

METRIC SYSTEM

Measures of Weight

(The Unit is the Gram = 0.035274 Ounces)

1 Milligram	=		=	0.015432 Grains
1 Centigram	=	10 Milligrams	=	0.15432 Grains
1 Decigram	=	10 Centigrams	=	1.5432 Grains
1 Gram	=	10 Decigrams	=	15.4323 Grains
1 Dekagram	=	10 Grams	=	5.6438 Drams
1 Hectogram	=	10 Dekagrams	=	3.5274 Ounces
1 Kilogram	=	10 Hectograms	=	2.2046223 Pounds
1 Myriagram	=	10 Kilograms	=	22.046223 Pounds
1 Quintal	=	10 Myriagrams	=	1.986412 Cwt.
1 Metric Ton	=	10 Quintal	=	22045.622 Pounds
1 Gram	=	0.56438 Drams		
1 Dram	=	1.77186 Grams		
	=	27.3438 Grains		
1 Metric Ton	=	2204.6223 Pounds		

 **METRIC SYSTEM**

Measures of Capacity

(The Unit is the Liter = 1.0567 Liquid Quarts)

1 Centiliter	= 10 Milliliters	=	0.338	Fluid Ounces
1 Deciliter	= 10 Centiliters	=	3.38	Fluid Ounces
1 Liter	= 10 Deciliters	=	33.8	Fluid Ounces
1 Dekaliter	= 10 Liters	=	0.284	Bushel
1 Hectoliter	= 10 Dekaliters	=	2.84	Bushels
1 Kiloliter	= 10 Hectoliters	=	264.2	Gallons

Note: $\dfrac{Kilometers}{8} \times 5 = Miles$ *or* $\dfrac{Miles}{5} \times 8 = Kilometers$

METRIC DESIGNATOR AND TRADE SIZES

Metric Designator												
12	16	21	27	35	41	53	63	78	91	103	129	155
3/8	1/2	3/4	1	1 1/4	1 1/2	2	2 1/2	3	3 1/2	4	5	6

TRADE SIZE

U.S. Weights and Measures/Metric Equivalent Chart

	In.	Ft.	Yd.	Mile	mm	cm	m	km
1 Inch =	1	.0833	.0278	1.578×10^{-5}	25.4	2.54	.0254	2.54×10^{-5}
1 Foot =	12	1	.333	1.894×10^{-4}	304.8	30.48	.3048	3.048×10^{-4}
1 Yard =	36	3	1	5.6818×10^{-4}	914.4	91.44	.9144	9.144×10^{-4}
1 Mile =	63360	5280	1760	1	1609344	160934.4	1609.344	1.609344
1 mm =	.03937	.0032808	1.0936×10^{-3}	6.2137×10^{-7}	1	0.1	0.001	0.000001
1 cm =	.3937	.0328084	.0109361	6.2137×10^{-6}	10	1	0.01	0.00001
1 m =	39.37	3.28084	1.09361	6.2137×10^{-4}	1000	100	1	0.001
1 km =	39370	3280.84	1093.61	0.62137	1000000	100000	1000	1

In. = Inches   Ft. = Foot   Yd. = Yard   Mi. = Mile   mm = Millimeter   cm = Centimeter   m = Meter   km = Kilometer

Explanation of Scientific Notation

Scientific Notation is simply a way of expressing very large or very small numbers in a more compact format. Any number can be expressed as a number between 1 and 10, multiplied by a power of 10 (which indicates the correct position of the decimal point in the original number). Numbers greater than 10 have positive powers of 10, and numbers less than 1 have negative powers of 10.

Example: $186000 = 1.86 \times 10^5$ $0.000524 = 5.24 \times 10^{-4}$

METRIC DESIGNATOR AND TRADE SIZES

Useful Conversions/Equivalents

1	BTU	Raises 1 lb of water 1°F
1	Gram Calorie	Raises 1 gram of water 1°C
1	Circular Mil	Equals 0.7854 sq. mil
1	SQ. Mil.	Equals 1.27 cir. mils
1	Mil	Equals 0.001 in.

To determine circular mil of a conductor:

Round Conductor . . . . CM = $(\text{Diameter in mils})^2$

Bus Bar . . . . CM = $\dfrac{\text{Width (mils)} \times \text{Thickness (mils)}}{0.7854}$

Notes:
1 Millimeter = 39.37 Mils
1 Cir. Millimeter = 1550 Cir. Mils
1 Sq. Millimeter = 1974 Cir. Mils

Note: The term *thousands of circular mills* was formerly abbreviated as *MCM* and is now abbreviated as *kcmil.*

DECIMAL EQUIVALENTS

Fraction					Decimal
1/64					.0156
2/64	1/32				.0313
3/64					.0469
4/64	2/32	1/16			.0625
5/64					.0781
6/64	3/32				.0938
7/64					.1094
8/64	4/32	2/16	1/8		.125
9/64					.1406
10/64	5/32				.1563
11/64					.1719
12/64	6/32	3/16			.1875
13/64					.2031
14/64	7/32				.2188
15/64					.2344
16/64	8/32	4/16	2/8	1/4	.25
17/64					.2656
18/64	9/32				.2813
19/64					.2969
20/64	10/32	5/16			.3125
21/64					.3281
22/64	11/32				.3438
23/64					.3594
24/64	12/32	6/16	3/8		.375
25/64					.3906
26/64	13/32				.4063
27/64					.4219
28/64	14/32	7/16			.4375
29/64					.4531
30/64	15/32				.4688
31/64					.4844
32/64	16/32	8/16	4/8	2/4	.5

Decimals are rounded to the nearest 10000th.

(*continued on next page*)

DECIMAL EQUIVALENTS

Fraction					Decimal
33/64					.5156
34/64	17/32				.5313
35/64					.5469
36/64	18/32	9/16			.5625
37/64					.5781
38/64	19/32				.5938
39/64					.6094
40/64	20/32	10/16	5/8		.625
41/64					.6406
42/64	21/32				.6563
43/64					.6719
44/64	22/32	11/16			.6875
45/64					.7031
46/64	23/32				.7188
47/64					.7344
48/64	24/32	12/16	6/8	3/4	.75
49/64					.7656
50/64	25/32				.7813
51/64					.7969
52/64	26/32	13/16			.8125
53/64					.8281
54/64	27/32				.8438
55/64					.8594
56/64	28/32	14/16	7/8		.875
57/64					.8906
58/64	29/32				.9063
59/64					.9219
60/64	30/32	15/16			.9375
61/64					.9531
62/64	31/32				.9688
63/64					.9844
64/64	32/32	16/16	8/8	4/4	1.000

Decimals are rounded to the nearest 10000th.

 **SINGLE-PHASE MOTORS**

Split-Phase—Squirrel-Cage—Dual-Voltage

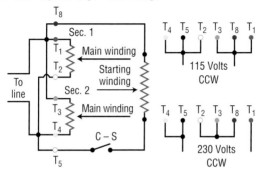

115 Volts
CCW

230 Volts
CCW

To Reverse,
Interchange 5 and 8

Classes of Single-Phase Motors:
1. Split-Phase
   A. Capacitor-Start
   B. Repulsion-Start
   C. Resistance-Start
   D. Split-Capacitor

2. Commutator
   A. Repulsion
   B. Series

Terminal Color Marking:

T_1 <u>Blue</u> • T_3 <u>Orange</u> • T_5 <u>Black</u> •

T_2 <u>White</u> T_4 <u>Yellow</u> ◦ T_8 <u>Red</u> •

Note: Split-phase motors are usually fractional horsepower. The majority of electric motors used in washing machines, refrigerators, etc. are of the split-phase type.

To change the speed of a split-phase motor, the number of poles must be changed.

1. Addition of running winding
2. Two starting windings, and two running windings
3. Consequent pole connections

SINGLE-PHASE MOTORS

Split-Phase—Squirrel-Cage

A. Resistance Start:

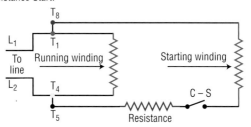

Centrifugal switch (CS) opens after reaching 75% of normal speed.

B. Capacitor Start:

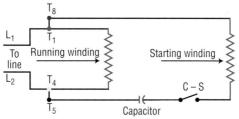

Notes:
1. A resistance start motor has a resistance connected in series with the starting winding.
2. The capacitor start motor is employed where a high starting torque is required.

CAPACITOR START MOTOR CIRCUIT

When Starting

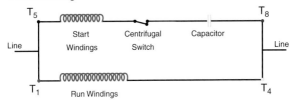

T_5 T_8

Start Windings   Centrifugal Switch   Capacitor

Line   Line

T_1 Run Windings T_4

When Running

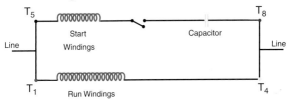

T_5 T_8

Start Windings   Capacitor

Line   Line

T_1 Run Windings T_4

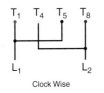

T_1 T_4 T_5 T_8

L_1 L_2

Clock Wise

T_1 T_4 T_5 T_8

L_1 L_2

DIRECT-CURRENT MOTORS

Terminal Markings

Terminal markings are used to tag terminals to which connections are to be made from outside circuits.

Facing the end opposite the drive (commutator end) the standard direction of shaft rotation is counter clockwise.

A-1 and A-2 indicate armature leads.
S-1 and S-2 indicate series-field leads.
F-1 and F-2 indicate shunt-field leads.

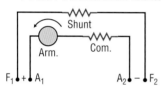

Shunt-Wound Motors
To change rotation, reverse either armature leads or shunt leads. **Do not** reverse both armature and shunt leads.

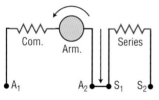

Series-Wound Motors
To change rotation, reverse either armature leads or series leads. **Do not** reverse both armature and series leads.

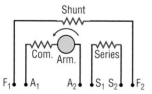

Compound-Wound Motors
To change rotation, reverse either armature leads or both the series and shunt leads. **Do not** reverse all three sets of leads.

NOTE: Standard rotation for DC generator is clockwise.

MOTOR SELECTION CHECKLIST

1. Horsepower requirements

2. Torque requirements

3. Speed requirements

4. Position (vertical, horizontal, etc.)

5. Conditions (temperature, water, corrosion, dust, etc.)

6. Operating cycle

7. Direction of rotation

8. Endplay

9. Available voltage, phases, frequency

10. Available starting current

11. Power factor concerns

MOTOR SELECTION CRITERIA

1. **Enclosures.** Must be suitable to area of installation, but enclosed motors are more expensive for both purchase and operation.

2. **Torque.** Motor's torque must exceed maximum torque requirements of the load. High-slip motors are preferred where there will be frequent peaks in required torque.

3. **Load cycle.** If loads cycle regularly, an average load (by RMS method) may be assumed, with a safety margin.

4. **Loading.** Motors operate most efficiently when fully loaded.

5. **Ambient temperatures.** Motors are designed for an ambient operating temperature of 40°C. Each 10° above this will halve the life of types A and B winding insulation.

TYPICAL LOAD SERVICE FACTORS

Load	Service Factor
Pump—centrifugal	1.0
Pump—centrifugal, sewage	2.0
Pump—rotary	1.5
Pump—reciprocating	2.0
Fan—light-duty	1.0
Fan—centrifugal	1.5
Blower—centrifugal	1.0
Blower—vane	1.25
Compressor—centrifugal	1.25
Compressor—vane	1.5
Elevator—bucket	2.0
Elevator—freight	2.25
Conveyor	1.5
Conveyor—heavy use	2.0
Punch press	2.25
Extruder—plastic	2.0
Extruder—metal	2.5
Concrete mixer	2.0
Printing press	1.5
Woodworking machines	1.0

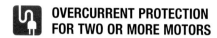 **OVERCURRENT PROTECTION FOR TWO OR MORE MOTORS**

1. Determine the size of an overcurrent protection device, sized for the largest motor.

2. Add size determined in step 1 to the full-load amperage of all other motors.

3. The group overcurrent device for the motor feeder may be no larger than the amperage arrived at in step 2.

See *NEC* sections:

- 430.62(A) on overcurrent devices

- 430.24 on feeders

- 366.22 on gutters (if necessary)

OVERCURRENT PROTECTION FOR MOTORS AND OTHER LOADS

1. Determine the size of an overcurrent protection device, based upon the largest motor.

2. Add the sum of all other loads to size determined in step 1.

3. The group overcurrent device may be no larger than the amperage arrived at in step 2.

See *NEC* sections:

- 430.62(A) on overcurrent devices

- 215.2(A)(1) on feeders

- 250.142(B) on sub-panels

- 240.6(A) on overcurrent device sizes

DETERMINING OVERLOAD SIZE

1. Find motor full-load amps.

2. Add 25% (to achieve 125% total).

3. Choose overloads based upon step 2.

Example: If full-load current is 60 amps, overloads must be chosen based upon 60 × 1.25, or 75 amps.

See *NEC* sections:

- 430.6(A)(2) on full-load currents

- 430.32(A)(1) on adding 25%

DETERMINING CONTROLLER SIZE

1. Determine horsepower size of motor.

2. Controller may have a horsepower rating no less than horsepower size of motor.

See *NEC* sections:

- 430.83(A)(1) on controllers

- 430.110(A) on disconnecting means

MOTORS 2 HORSEPOWER OR LESS AND 300 VOLTS OR LESS

Switches may be used as controllers provided:

1. A general use switch can be used if it is rated for at least twice the full-load current of the motor.

2. AC switches on AC circuits may be used if the switch is rated 125% of the full-load current of the motor.

See *NEC* sections:

- 430.83(C)(1) on general use switches

- 430.83(C)(2) on AC switches

DETERMINING CONDUCTOR SIZES FOR SINGLE-PHASE MOTORS

1. Find motor full-load amps.

2. Add 25% to full-load amps (to achieve 125%).

3. Conductor must have an ampacity no lower than determined in step 2.

4. Select conductors from *NEC* Table 310.16.

See *NEC* sections:

* 430.6(A)(1) on full-load currents

* 430.22(A) on adding 25%

 **DETERMINING CONDUCTOR SIZES FOR ADJUSTABLE SPEED DRIVES**

1. Determine rated input of power conversion equipment.

2. Add 25% (to achieve 125%) of rated current.

3. Select conductors from *NEC* Table 310.15(B)(16).

See *NEC* sections:

- 430.122(A) on adjustable speed drive conductors
- 430.2 on variable speed motors

 MOTOR AND MOTOR-CIRCUIT CONDUCTOR PROTECTION

Motors can have starting currents three to five times (or more) than that of the motor's normal current. To allow such motors to start, the motor circuit conductors are allowed to be protected by circuit breakers and fuses at values that are higher than the actual motor and conductor ampere ratings. These larger overcurrent devices do not provide full overload protection and will open only when exposed to larger overcurrents, such as those associated with short circuits or ground faults. Overload protection (based on the actual nameplate amperes of the motor) must be used to protect the installation. This protection is usually in the form of heating elements in manual or magnetic motor starters. Small motors such as waste disposal motors have a red overload reset button built into the motor.

GENERAL MOTOR RULES

- Use Full-Load Current from tables instead of nameplate.

- Branch-Circuit Conductors: Use 125% of Full-Load Current to find conductor size.

- Branch-Circuit OCP Size: Use percentages given in tables for Full-Load Current.

- Feeder Conductor Size: 125% of largest motor and sum of the rest.

- Feeder OCP: Use largest OCP plus rest of Full-Load Currents.

 MOTOR BRANCH CIRCUIT AND FEEDER EXAMPLE

General Motor Applications

Branch-Circuit Conductors: Use Full-Load Three-Phase Currents; From *NEC* Table 430.250,
50 HP 480 Volt Three-Phase motor design B, 75 degree terminations = 65 Amperes
125% of Full-Load Current [*NEC* 430.22(A)]
125% of 65 A = **81.25 Amperes** Conductor Selection Ampacity

Branch-Circuit Overcurrent Device: *NEC* 430.52 (C)(1)
(Branch-Circuit Short Circuit and Ground-Fault Protection)
Use percentages given in *NEC* 430.52 for **Type**
of circuit breaker or fuse used.
50 HP 480 V 3 Ph Motor = 65 Amperes.
Nontime Fuse = 300%.
300% of 65A = 195 A. *NEC* 430.52(C)(1)(EX1) Next size allowed
NEC 240. 6A = **200 Ampere Fuse**.

**MOTOR BRANCH CIRCUIT AND
FEEDER EXAMPLE**

Feeder Connectors: For 50 HP and 30 HP 480 Volt Three-Phase
design B motors on same feeder
Use 125% of largest full-load current and 100% of rest. (*NEC* 430.24)
50 HP 480 V 3 Ph Motor = 65A; 30 HP 480 V 3 Ph Motor = 40A
(125% of 65A) + 40A = **121.25 A** Conductor Selection Ampacity

Feeder Overcurrent Device: [*NEC* 430.62(A)]
(Feeder short circuit and ground-fault protection)
Use largest overcurrent protection device <u>plus</u> full-load currents of
the rest of the motors.
50 HP = 200 A fuse   (65 FLC)
30 HP = 125 A fuse   (40 FLC)
200 A fuse + 40 A (FLC) = 240 A. Do not exceed this value on feeder.
Go down to a **225 A** fuse.

APPROXIMATE TORQUE FIGURES, COMPOUND DC MOTORS

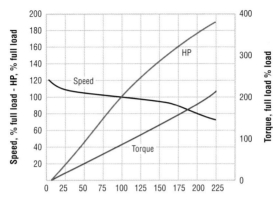

Speed and torque are inversely proportional; as one rises, the other falls.

HEATER CORRECTIONS FOR AMBIENT TEMPERATURES

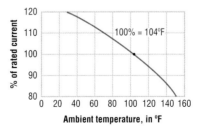

100% is standardized at 104°F (40°C). Electrical components must be protected from excess heat.

87

APPROXIMATE TORQUE FIGURES, AC MOTORS

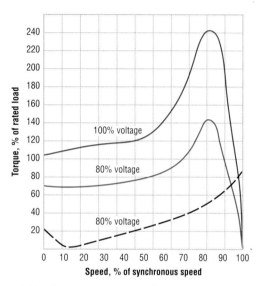

A standard AC motor may never reach synchronous speed. As it approaches synchronous speed, torque falls to zero and rotation slows.

APPROXIMATE TORQUE FIGURES, WOUND ROTOR MOTORS

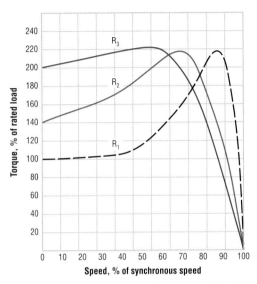

Torque peaks at higher percentages of synchronous speed, but falls to zero as synchronous speed is approached.

SIZING LOAD CONDUCTORS FROM GENERATORS

1. Determine generator output current from nameplate.

2. Add 15% (to achieve 115%) to generator current.

3. Select conductors from *NEC* Table 310.15(B)(16).

See *NEC* sections:

- 445.13 on conductor ampacity

- 445.13, exception, on an exception to the 15% adder

ELECTRICAL SAFETY DEFINITIONS

Arc—An electrical discharge through air, commonly called a "spark." Arcs occur when the voltage between two conductive surfaces breaks down the gasses in between, turning it into a conductive plasma.

Arc Blast—A pressure wave due to the heating, melting, vaporization, and expansion of conductor material and the surrounding air. Sometimes also called a "splash."

Arc Flash—The release of heat and light at the point of an arc. Properly, arcs (see above) are discharges through plasma, but it is common among electricians to refer to violent contact between conductors (as in the case of an accidental dead short) as an arc.

Approach Boundaries—Boundaries established to protect personnel from electrical shock and associated hazards.

Calorie—The amount of heat required to raise the temperature of 1 gram of water by 1° Celsius.

Electrically Safe Work Condition—A condition in which equipment and/or circuit components have been disconnected from sources of electrical energy, locked/tagged out, and tested to verify that all sources of power are disconnected.

For more information, refer to *NFPA 70E, Standard for Electrical Safety in the Workplace.*

Reprinted with permission from Littlefuse®, *www.littlefuse.com*; 1-800-TEC-FUSE

ELECTRICAL SAFETY DEFINITIONS

Exposed Live Parts—An energized conductor or object that is capable of being inadvertently touched or approached (nearer than a safe distance) by a person or equipment that could become energized. It is applicable to parts that are not in an electrically safe work condition; that is, suitably grounded, isolated, or insulated.

Flame Resistant—The chemical characteristic of fabric (or similar materials) that is designed to limit the ignition or burning. The term can be applied to specific characteristics of the material or a treatment applied to a material.

Flash Hazard Analysis—An analysis of possible exposure to Arc-Flash hazards. Such an analysis specifies Incident Energy levels, Hazard Risk Categories, Flash Protection Boundaries, and required personal protective equipment (PPE). It may also help define safe work practices.

Flash Protection Boundary—A boundary established to protect people and equipment from Arc-Flash hazards. The Flash Protection Boundary is the distance at which an unprotected worker might receive a second-degree burn to bare skin.

Flash Suit—Fire-resistant rated Personal Protective Equipment, also referred to as "FR rated PPE." A combined set of pants, shirt/jacket, and flash hood with a built-in face shield system. (Hand and foot protection are normally not included.)

Hazard Risk Category—A classification of risks, generally ranging between 1 and 4, defined by NFPA 70E. Each category comes with specified energy levels and personal protective equipment (PPE).

For more information, refer to *NFPA 70E, Standard for Electrical Safety in the Workplace.*

Reprinted with permission from Littlefuse®, *www.littlefuse.com*; 1-800-TEC-FUSE

ELECTRICAL SAFETY DEFINITIONS

Incident Energy—The amount of thermal energy to which a surface (at a given distance) is exposed during an electrical arc at a certain distance from the arc. Incident energy is typically measured in calories per square centimeter (cal/cm^2).

PPE—An acronym for Personal Protective Equipment. PPE can include clothing, tools, or equipment.

Qualified Person—A person who is considered to be trained and knowledgeable in the construction and operation of equipment, and who is capable of recognizing electrical hazards that may be encountered.

Shock, electrical shock—A physical trauma caused by electrical current. Electrical shocks can be lower than the threshold of human perception, or large enough to cause instant death. The effects of electrical shock depend on the amount of current running through a body, as well as the path of the current through the body.

Unqualified Person—A person who is not considered to be trained and knowledgeable in the construction and operation of equipment, and who is not capable of recognizing electrical hazards that may be encountered.

For more information, refer to *NFPA 70E, Standard for Electrical Safety in the Workplace.*

Reprinted with permission from Littlefuse®, *www.littlefuse.com*; 1-800-TEC-FUSE

ELECTRICAL SAFETY CHECKLIST

1. Deenergize equipment whenever practicable prior to performing any work on it.

2. Verify that you understand the situation you are facing. Get outside input as required.

3. Verify that you know how to address all risks.

4. Verify that you have the necessary equipment and support to address the situation.

5. If you plan to work on energized equipment ("working hot"), stop and reassess to be sure that it is the right choice.

6. Specify how you will ensure safety.

7. Determine (if necessary) whether or not a Hazard Analysis has been performed to identify hazards such as shock, arc flash, etc.

8. Identify (as required) protection boundaries for shock (Limited & Restricted Approach) and arc flash (Flash Protection Boundary).

9. Identify the required Personal Protective Equipment (PPE) for the task to be performed. Also consult a Hazard Risk Category (HRC) analysis, if available.

10. Provide barriers or other means to prevent access to the work area by anyone lacking the knowledge and skills that are necessary.

11. If other people are involved, call a job briefing and identify job- or task-specific hazards.

12. Where required, obtain written management approval to perform work on energized equipment.

For more information, refer to *NFPA 70E, Standard for Electrical Safety in the Workplace.*

Reprinted with permission from Littlefuse®, *www.littlefuse.com*; 1-800-TEC-FUSE

ELECTRICAL SAFETY
LOCKOUT–TAGOUT PROCEDURES

Some sections of the OSHA code require the energy sources to machines or equipment to be turned off and disconnected, isolating them from the energy source, before work is performed. Such isolating or disconnecting means must be either locked or tagged with a warning label. While lockout is the more reliable and preferred method, OSHA accepts tagout to be a suitable replacement in some situations. NFPA standard 70E (in Article 120) contains detailed instructions for lockout–tagout procedures. Its goal is "an electrically safe work condition."

Application of Lockout–Tagout Devices

1. Make all necessary preparations for shutdown.

2. Shut down the machine or equipment in question.

3. Deenergize the circuit(s) in question at all appropriate overcurrent or isolating devices (usually a fuse or circuit breaker).

4. Install the lockout and/or tagout device(s).

5. Be careful to drain energized capacitors and other sources of residual energy.

6. Test to verify that all machines and/or equipment are isolated and deenergized, and ensure that all affected people are aware of the work and the deenergization.

For more information, refer to *NFPA 70E, Standard for Electrical Safety in the Workplace.*

Reprinted with permission from Littelfuse®, *www.littelfuse.com*; 1-800-TEC-FUSE

ELECTRICAL SAFETY
LOCKOUT–TAGOUT PROCEDURES

Removal of Lockout–Tagout Devices

1. Verify that no people are in locations that might subject them to electrical shock or hazards arising from the re-energizing of the systems in question.

2. Make sure that no tools, fasteners, or other parts have been left in places where they could constitute a hazard.

3. Inspect the work area to ensure that the machine and/or equipment components are intact and capable of operating properly.

4. Verify that all affected people are aware of the re-energization.

5. After removing locks or tags, perform one final check before restarting the equipment or machines.

Note: For specific lockout–tagout procedures, refer to OSHA and NFPA 70E.

For more information, refer to *NFPA 70E, Standard for Electrical Safety in the Workplace.*

Reprinted with permission from Littlefuse®, *www.littlefuse.com*; 1-800-TEC-FUSE

ELECTRICAL SAFETY SHOCK PROTECTION BOUNDARIES

Nominal System Voltage (Phase to Phase)	Limited Approach Exposed Fixed Circuit Part	Boundary Exposed Movable Conductor	Restricted Approach Boundary
50 to 300 V	10 ft. 0 in.	3 ft. 6 in.	Avoid Contact
301 to 750 V	10 ft. 0 in.	3 ft. 6 in.	1 ft. 0 in.
751 V to 15 kV	10 ft. 0 in.	5 ft. 0 in.	2 ft. 2 in.
15.1 kV to 36 kV	10 ft. 0 in.	6 ft. 0 in.	2 ft. 7 in.
36.1 kV to 46 kV	10 ft. 0 in.	8 ft. 0 in.	2 ft. 9 in.
46.1 kV to 72.5 kV	10 ft. 0 in.	8 ft. 0 in.	3 ft. 3 in.
72.6 kV to 121 kV	10 ft. 8 in.	8 ft. 0 in.	3 ft. 4 in.

Note: Data derived from NFPA 70E Table 130.2(C).

Shock protection boundaries are based on system voltages, as well as the fixed/movable characteristics of the exposed energized components. NFPA 70E Table 130.2(C) defines these boundary distances for nominal phase-to-phase system voltages from 50 Volts to 800 kV. Approach Boundary distances may range from an inch to several feet. Refer to NFPA 70E Table 130.2(C) for more information.

Protection Boundaries

Limited Approach: A qualified person, or unqualified person if accompanied by qualified person. PPE is required.

Restricted Approach: Qualified persons only. PPE is required.

For more information, refer to *NFPA 70E, Standard for Electrical Safety in the Workplace.*

Reprinted with permission from Littlefuse®, *www.littlefuse.com*; 1-800-TEC-FUSE

ELECTRICAL SAFETY
HOW TO READ A WARNING LABEL

The amount of heat energy (cal/cm²) at the distance shown. The incident energy determines the Hazard Risk Category.

The distance from exposed energized parts at which a 2nd degree burn can occur to unprotected skin.

The *NFPA* 70E established Hazard Risk Category (see *NFPA 70E* Table 130.7(c)(11) for explanation) based on Incident Energy.

⚠ WARNING

Arc-Flash and Shock Hazard

Appropriate PPE Required

ARC-FLASH PROTECTION BOUNDARY AND REQUIRED PPE

Flash Hazard Boundary	41 inches	Hazard Risk Category	2
Incident Energy at 18" (cal/cm²)	4.05 cal/cm²	Glove Class	00
Required PPE	Cotton Underwear + FR Shirt & Pants + Safety Glasses + Hard Hat + Leather Gloves & Shoes + Ear Plugs + Face Shield		

SHOCK HAZARD PROTECTION BOUNDARIES

Shock Hazard	480 VAC				
Limited	42 inches	Restricted	12 inches	Prohibited	1 inch
Equipment ID:	Panel L-10	Assessment Date:	8/03/07		

Littelfuse
Expertise Applied | Answers Delivered

800-TEC-FUSE
www.littelfuse.com

Equipment voltage determining the shock approach boundaries.

Name or ID of specific electrical equipment for which this label is produced.

Required PPE (personal protective equipment) based on the Incident Energy and Hazard Risk Category.

Required glove class to protect against voltage and shock hazard.

For more information, refer to *NFPA 70E, Standard for Electrical Safety in the Workplace.*

Reprinted with permission from Little fuse®, *www.Little fuse.com;* 1-800-TEC-FUSE

MOTOR MAINTENANCE—ANNUAL

Designation _____   Serial No. _____

Location _____   Date Installed _____

Type _____ Frame _____   Manufacturer _____

Voltage _____ HP _____   RPM _____

✓	Process	Initials
	Lock-out power to motor and control circuit	
	Exterior cleaning	
	Uncouple motor from its load, etc.	
	Clear inside of motor	
	Check motor, windings, switch assemblies	
	Check bearings	
	Check slip rings, brushes, commutator, etc.	
	Reassemble motor and load, check drive mechanism	
	Flush and replace bearing lubricant	
	Check raceway, terminations, capacitor, etc.	
	Check mounting, bolts, etc.	
	Optional: Megohmmeter testing	
	Inspect controls and circuits	
	Reconnect power, check all voltages	
	Run motor, check current with nameplate	
	Check bearings	

MOTOR MAINTENANCE—SEMI-ANNUAL

Designation _____   Serial No. _____

Location_____   Date Installed _____

Type_____ Frame _____   Manufacturer _____

Voltage_____ HP _____   RPM_____

✓	Process	Initials
	Lock-out power to motor and control circuit	
	Exterior cleaning	
	Check raceways and terminations	
	Check drive mechanisms	
	Lubricate bearings as required	
	Check slip rings, brushes, commutators, etc.	
	Check mounting, bolts, etc.	
	Inspect controls and circuits	
	Reconnect power, check all voltages	
	Run motor, final check	

PULLEY CALCULATIONS

The most common configuration consists of a motor with a pulley attached to its shaft, connected by a belt to a second pulley. The motor pulley is referred to as the **Driving Pulley**. The second pulley is called the **Driven Pulley**. The speed at which the Driven Pulley turns is determined by the speed at which the Driving Pulley turns as well as the diameters of both pulleys. The following formulas may be used to determine the relationships between the motor, pulley diameters and pulley speeds.

D = **Diameter of Driving Pulley**
d^1 = **Diameter of Driven Pulley**
S = **Speed of Driving Pulley** (revolutions per minute)
s^1 = **Speed of Driven Pulley** (revolutions per minute)

Driving Pulley   Driven Pulley

- *To determine the speed of the Driven Pulley (Driven RPM):*

$$s^1 = \frac{D \times S}{d^1} \quad \text{or} \quad \text{Driven RPM} = \frac{\text{Driving Pulley Dia.} \times \text{Driving RPM}}{\text{Driven Pulley Dia.}}$$

- *To determine the speed of the Driving Pulley (Driving RPM):*

$$s = \frac{d^1 \times s^1}{D} \quad \text{or} \quad \text{Driving RPM} = \frac{\text{Driven Pulley Dia.} \times \text{Driven RPM}}{\text{Driving Pulley Dia.}}$$

- *To determine the diameter of the Driven Pulley (Driven Dia.):*

$$d^1 = \frac{D \times S}{s^1} \quad \text{or} \quad \text{Driven Dia.} = \frac{\text{Driving Pulley Dia.} \times \text{Driving RPM}}{\text{Driven RPM}}$$

- *To determine the diameter of the Driving Pulley (Driving Dia.):*

$$D = \frac{d^1 \times s^1}{S} \quad \text{or} \quad \text{Driving Dia.} = \frac{\text{Driven Pulley Dia.} \times \text{Driven RPM}}{\text{Driving RPM}}$$

DETERMINING BELT LENGTH

$$\text{Length} = \frac{\pi(D+d)}{2} + \sqrt{x^2 + (\frac{D-d}{2})^2}$$

D = Diameter of larger pulley
d = Diameter of smaller pulley
$\pi = 3.1416$
x = Distance between shaft centers

 HORSEPOWER CAPACITIES

Belt Speed (ft./min)	Pulley Diameter						
	1/2"	1"	1½"	2"	3"	4"	6"
1000	.33	.53	.66	.79	.92	.99	1.05
2000	.99	1.32	1.38	1.45	1.65	1.78	1.91
3000	1.25	1.58	1.98	2.17	2.59	2.70	2.90
4000	1.51	1.91	2.44	2.77	3.17	3.43	3.76
5000	1.65	2.11	2.77	3.10	3.63	3.96	4.42
6000	1.71	2.31	3.03	3.36	3.96	4.42	4.95
7000	1.71	2.37	3.16	3.56	4.29	4.88	5.54
8000	1.65	2.44	3.30	3.69	4.49	4.55	5.87
9000	1.51	2.50	3.36	3.76	4.55	4.62	6.07
10000	1.32	2.50	3.43	3.82	4.62	4.62	6.14

Notes:
Based on medium, Single-Ply, Dacron Belts
per inch of width

GEAR SIZING

$$N = \frac{n \times R}{r}$$

$$n = \frac{N \times R}{r}$$

n = Number of teeth, driven gear
N = Number of teeth, driving gear
R = RPM, pinion
r = RPM, gear

 DETERMINING SHAFT DIAMETER

Shaft diameter (in inches) = $\sqrt{\dfrac{K \times HP}{RPM}}$

HP = Horsepower transmitted
K = Constant, varying between 50 and 125, depending on shaft and distance between bearings

GEAR REDUCERS

Output Torque

$$O_T = I_T \times R_R \times R_E$$

Output Speed

$$O_S = \frac{I_S}{R_R} \times R_E$$

Output Horsepower

$$O_{HP} = I_{HP} \times R_E$$

O_T = Output torque
I_T = Input torque
R_R = Gear reduction ratio
O_S = Output speed (RPM)
I_S = Input speed (RPM)
R_E = Reducer efficiency
O_{HP} = Output horsepower
I_{HP} = Input horsepower

MOTOR TORQUE

Torque

$$T = \frac{HP \times 5252}{RPM}$$

Starting Torque

$$T = \frac{HP \times 5252 \times C}{RPM}$$

T   = Torque, 1 lb-ft
HP = Horsepower
RPM = Rotations per minute
C   = Motor class percentage

5252 is a constant, derived as follows: $\dfrac{33000 \; 1 \; \text{lb-ft}}{\pi \times 2} = 5252$

CALCULATING COST OF OPERATING AN ELECTRICAL APPLIANCE

What is the monthly cost of operating a 240 volt 5 kilowatt (kW) central electric heater that operates 12 hours per day, when the cost is 15 cents per kilowatt-hour (kWhr)?

Cost = Watts x Hours used x Rate per kWhr / 1000

5 kW = 5000 Watts
Hours = 12 hours x 30 days = 360 hours per month

= 5000 x 360 x .15 / 1000
= 270000 / 1000 = **$270 Monthly cost**

The above example is for a resistive load. Air-conditioning loads are primarily inductive loads. However, if ampere and voltage values are known, this method will give an approximate cost. Kilowatt-hour rates vary for different power companies, and for residential use, graduated rate scales are usually used (the more power used, the lower the rate). Commercial and industrial rates are generally based on kilowatt usage, maximum demand, and power factor. Other costs are often added such as fuel cost adjustments.

ELECTRICAL SYMBOLS

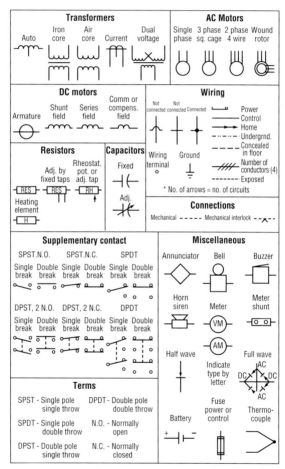

Transformers

Auto | Iron core | Air core | Current | Dual voltage

AC Motors

Single phase | 3 phase sq. cage | 2 phase 4 wire | Wound rotor

DC motors

Armature | Shunt field | Series field | Comm or compens. field

Wiring

Not connected | Not connected | Connected | Power
Home
Undergrnd.
Concealed in floor
Number of conductors (4)
Exposed

Wiring terminal | Ground

* No. of arrows = no. of circuits

Resistors

Adj. by fixed taps | Rheostat. pot. or adj. tap

RES | RES | RH

Heating element

H

Capacitors

Fixed

Adj.

Connections

Mechanical ---- Mechanical interlock

Supplementary contact

SPST.N.O. | SPST.N.C. | SPDT

Single break | Double break | Single break | Double break | Single break | Double break

DPST, 2 N.O. | DPST, 2 N.C. | DPDT

Single break | Double break | Single break | Double break | Single break | Double break

Miscellaneous

Annunciator | Bell | Buzzer

Horn siren | Meter | Meter shunt

VM

Half wave | AM | Full wave

Indicate type by letter

AC DC DC AC

Terms

SPST - Single pole single throw

SPDT - Single pole double throw

DPST - Double pole single throw

DPDT - Double pole double throw

N.O. - Normally open

N.C. - Normally closed

Battery

+ −

Fuse power or control

Thermo-couple

ELECTRICAL SYMBOLS

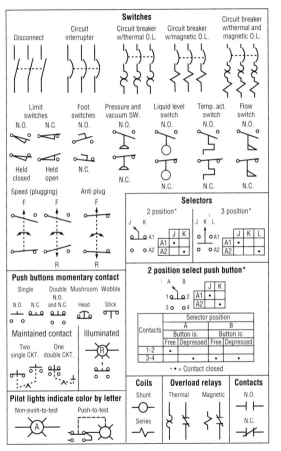

Switches

| Disconnect | Circuit interrupter | Circuit breaker w/thermal O.L. | Circuit breaker w/magnetic O.L. | Circuit breaker w/thermal and magnetic O.L. |

Limit switches	Foot switches	Pressure and vacuum SW.	Liquid level switch	Temp. act. switch	Flow switch
N.O. N.C.	N.O.	N.O.	N.O.	N.O.	N.O.
Held closed	Held open	N.C.			
			N.C.	N.C.	N.C.

Speed (plugging)   Anti plug

F   F   F

R   R

Selectors

2 position*   3 position*

2 position select push button*

Contacts	Selector position			
	A		B	
	Button is:		Button is:	
	Free	Depressed	Free	Depressed
1-2	•			
3-4		•		•

• • = Contact closed

Push buttons momentary contact

| Single | Double N.O. and N.C. | Mushroom Head | Wobble Stick |
| N.O. N.C. | | | |

Maintained contact   Illuminated

Two single CKT.   One double CKT.

Coils	Overload relays		Contacts
Shunt	Thermal	Magnetic	N.O.
Series			N.C.

Pilot lights indicate color by letter

Non-push-to-test   Push-to-test

Note: N.O. = Normally Open; N.C. = Normally Closed

107

 **WIRING DIAGRAMS**

Basic Diagram of Two-Wire Control Circuit

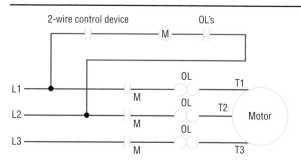

Wiring Diagram of Starter (Two-Wire Control)

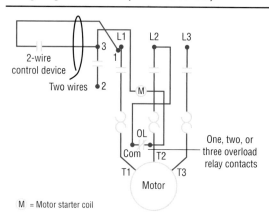

Ⓜ = Motor starter coil

Reprinted from Miller, Charles R. *NFPA's Pocket Electrical References, First Edition.* Jones and Bartlett Publishers.

WIRING DIAGRAMS

Control Circuit Only

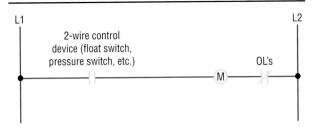

Basic Three-Wire Control Circuit

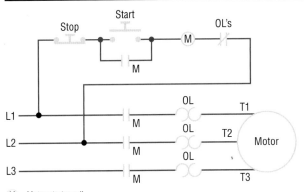

M = Motor starter coil

Reprinted from Miller, Charles R. *NFPA's Pocket Electrical References, First Edition.*
Jones and Bartlett Publishers.

WIRING DIAGRAMS

Wiring Diagram of Starter (Three-Wire Control)

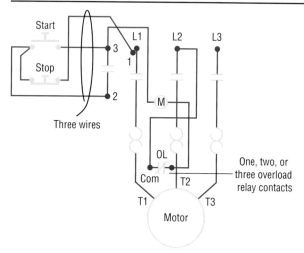

Control Circuit Only

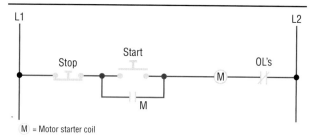

(M) = Motor starter coil

Reprinted from Miller, Charles R. *NFPA's Pocket Electrical References, First Edition.* Jones and Bartlett Publishers.

COMPLETE STOP-START SYSTEM WITH CONTROL TRANSFORMER

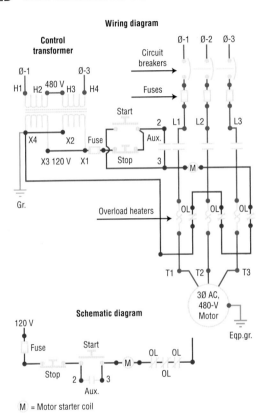

Wiring diagram

Control transformer

Ø-1 Ø-2 Ø-3

Circuit breakers

Fuses

H1 H2 480 V H3 H4

Start

X4 X2 Fuse 2 Aux. L1 L2 L3

X3 120 V X1 Stop 3 M

Gr.

Overload heaters

OL OL OL

T1 T2 T3

3Ø AC, 480-V Motor

Eqp.gr.

Schematic diagram

120 V

Fuse Start M OL OL

Stop 2 3 OL

Aux.

M = Motor starter coil

Note: Controls and motor are of different voltages.

HAND OFF AUTOMATIC CONTROL

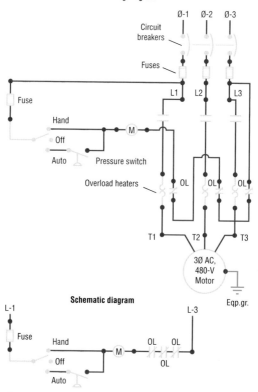

Wiring diagram

Ø-1 Ø-2 Ø-3

Circuit breakers

Fuses

L1 L2 L3

Fuse

Hand

Off

Auto

M

Pressure switch

Overload heaters

OL OL OL

T1 T2 T3

3Ø AC, 480-V Motor

Eqp.gr.

Schematic diagram

L-1

Fuse

Hand

Off

Auto

M

OL OL

OL

L-3

(M) = Motor starter coil

Note: Controls and motor are of the same voltage.
If Low Voltage controls are used, see page 111 for control transformer connections.

112

JOGGING WITH CONTROL RELAY

Jogging control

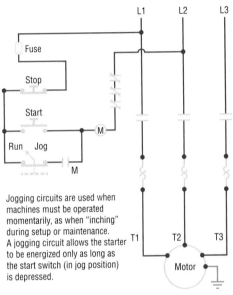

Jogging circuits are used when machines must be operated momentarily, as when "inching" during setup or maintenance. A jogging circuit allows the starter to be energized only as long as the start switch (in jog position) is depressed.

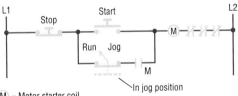

Ⓜ = Motor starter coil

Multiple Start and Stop Stations

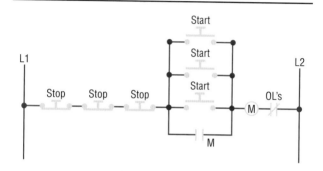

Start Push Button with Job Selector Switch

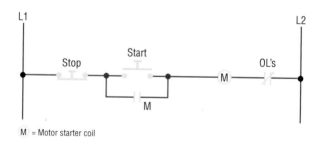

$\textcircled{M}$ = Motor starter coil

Reprinted from Miller, Charles R. *NFPA's Pocket Electrical References, First Edition.*
Jones and Bartlett Publishers.

114

WIRING DIAGRAMS

Reversing Starter with Limit Switches

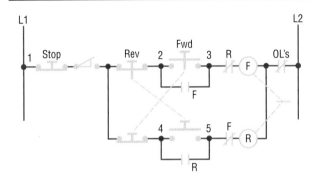

Reversing Starter

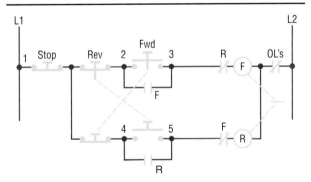

Reprinted from Miller, Charles R. *NFPA's Pocket Electrical References, First Edition.* Jones and Bartlett Publishers.

PLUGGING CIRCUIT

Plugging is a method of stopping a motor and load quickly by putting the motor into reverse. A centrifugal switch is necessary to prevent the motor from continuing in reverse.

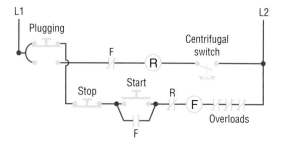

TERMINAL DESIGNATIONS

Generators and Synchronous Motors

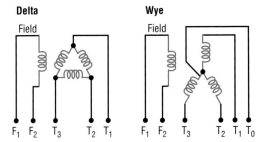

Delta

Field

F_1 F_2 T_3 T_2 T_1

Wye

Field

F_1 F_2 T_3 T_2 T_1 T_0

Terminal identifications are the same for both delta and wye, except that the wye arrangement includes a T_0 terminal, which connects to the center-point of the coils.

COUNTER-EMF STARTING

DC Motors

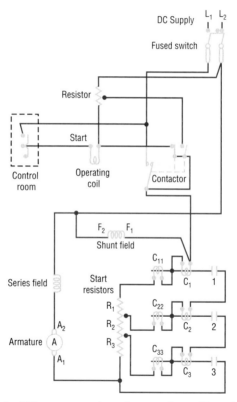

A counter-EMF opposes any change in current flow and is used to reduce high-starting currents.

TWO-SPEED STARTING

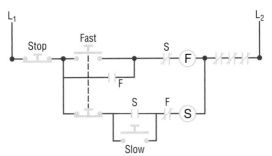

Note this circuit's interlocks:

- Closing the fast switch also opens the Slow leg of the circuit.

- When the S coil is activated, it opens the normally closed S contact in the Fast leg of the circuit.

- When the F coil is activated, it opens the normally closed F contact in the Slow leg of the circuit.

REDUCED-VOLTAGE STARTING

Synchronous Motor

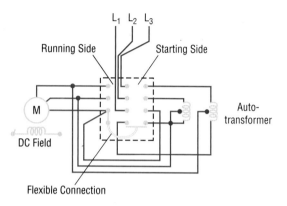

The auto-transformers are in the circuit for starting, and removed afterward.

CLICKSAFETY®
Safety. Compliance. Your Future.

The Marketplace
for Your Electrical
Safety Training Needs

ClickSafety offers:

- Accredited & Industry-Leading Electrical Safety Training Courses

- 100% Online: Learn at Your Own Pace

- FREE Electrical Safety Toolbox Talk Videos

- In-house Team of Multi-Accredited Safety & Health Professionals

Let Us Help Keep You Safe!
www.ClickSafety.com | sales@clicksafety.com
www.facebook.com/ClickSafety

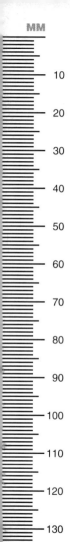

A note from the publisher . . .

We believe that *Ugly's Electric Motors and Controls* is the finest pocket electrical reference book for electric motors and controls available anywhere. It is our goal to continually improve this little yellow book so it retains its #1 position in the industry for years to come.

We welcome your comments.

If you have a suggestion on how we can make *Ugly's Electric Motors and Controls* a more valuable tool for you or your company, please write, phone, or fax us. We will seriously review all suggestions.

Jones & Bartlett Learning
5 Wall Street
Burlington, MA 01803
978-443-5000
info@jblearning.com
www.jblearning.com